Environmentally Conscious Alternative Energy Production

Environmentally Conscious Alternative Energy Production

Edited by
Myer Kutz

John Wiley & Sons, Inc.

Library of Congress Cataloging-in-Publication Data:

Environmentally conscious alternative energy production / edited by Myer Kutz.
    p. cm.
  Includes index.
  ISBN 978-0-471-73911-1 (cloth)
1.   Electric power production. 2.   Electric power production–Environmental aspects. 3.
  Renewable energy sources. 4.   Global warming–Prevention. 5.   Sustainable engineering.
  I. Kutz, Myer.
  TK1005.E58 2007
  621.31–dc22

                                                                    2007006006

Printed in the United States of America

10 9 8 7 6 5 4 3 2 1

To Bob and Linda, to Bob and Nadine, and to Linda

Contents

Contributors   ix

Preface   xi

1   Economic Comparisons of Power Generation Technologies   1
      Todd S. Nemec

2   Solar Energy Applications   13
      Jan F. Kreider

3   Fuel Cells   59
      Matthew M. Mench

4   Geothermal Resources and Technology: An Introduction   101
      Peter D. Blair

5   Wind Power Generation   119
      Todd S. Nemec

6   Cogeneration   129
      Jerald A. Caton

7   Hydrogen Energy   165
      E. K. Stefanakos, D. Y. Goswami, S. S. Srinivasan, and J. T. Wolan

8   Clean Power Generation from Coal   207
      James W. Butler and Prabir Basu

9   Using Waste Heat from Power Plants   267
      Herbert A. Ingley III

Appendix A   Solar Thermal and Photovoltaic Collector Manufacturing Activities 2005   275

Appendix B   Survey of Geothermal Heat Pump Shipments, 1990–2004   295

Index   297

Contributors

Prabir Basu
Department of Mechanical Engineering
Dalhousie University
Halifax, Nova Scotia
Canada

Peter D. Blair
National Academy of Sciences
Washington, DC

James W. Butler
Department of Mechanical Engineering
Dalhousie University
Halifax, Nova Scotia
Canada

Jerald A. Caton
Department of Mechanical Engineering
Texas A&M University
College Station, Texas

D. Y. Goswami
Clean Energy Research Center
University of South Florida
Tampa, Florida

Herbert A. Ingley III
University of Florida
Gainesville, Florida

Jan F. Kreider
Kreider and Associates, LLC
and
Joint Center for Energy
    Management
University of Colorado
Boulder, Colorado

Matthew M. Mench
Department of Mechanical and
    Nuclear Engineering
The Pennsylvania State University
University Park, Pennsylvania

Todd S. Nemec
GE Energy
Schenectady, New York

S. S. Srinivasan
Clean Energy Research Center
University of South Florida
Tampa, Florida

E. K. Stefanakos
Clean Energy Research Center
University of South Florida
Tampa, Florida

J. T. Wolan
Clean Energy Research Center
University of South Florida
Tampa, Florida

Preface

Many readers will approach the books in the **Wiley Series in Environmentally Conscious Engineering** with some degree of familiarity with, knowledge about, or even expertise in, one or more of a range of environmental issues, such as climate change, pollution, and waste. Such capabilities may be useful for readers of this series, but they aren't strictly necessary, for the purpose of this series is not to help engineering practitioners and managers deal with the *effects* of man-induced environmental change. Nor is it to argue about whether such effects degrade the environment only marginally or to such an extent that civilization, as we know it, is in peril, or that any effects are nothing more than a scientific-establishment-and-media-driven hoax and can be safely ignored. (Authors of a plethora of books, even including fiction, and an endless list of articles in scientific and technical journals, have weighed in on these matters, of course.) On the other hand, this series of engineering books does take as a given that the overwhelming majority in the scientific community is correct, and that the future of civilization depends on minimizing environmental damage from industrial, as well as personal, activities. At the same time, the series does not advocate solutions that emphasize only curtailing or cutting back on these activities. Instead, its purpose is to exhort and enable engineering practitioners and managers to reduce environmental impacts, to engage, in other words, in Environmentally Conscious Engineering, a catalog of practical technologies and techniques that can improve or modify just about anything engineers do, whether they are involved in designing something, making something, obtaining or manufacturing materials and chemicals with which to make something, generating power, or transporting people and freight.

Increasingly, engineering practitioners and managers need to know how to respond to challenges of integrating environmentally conscious technologies, techniques, strategies, and objectives into their daily work, and, thereby, find opportunities to lower costs and increase profits while managing to limit environmental impacts. Engineering practitioners and managers also increasingly face challenges in complying with changing environmental laws. So companies seeking a competitive advantage and better bottom lines are employing environmentally responsible design and production methods to meet the demands of their stakeholders, who now include not only owners and stockholders, but also customers, regulators, employees, and the larger, even worldwide community.

Engineering professionals need references that go far beyond traditional primers that cover only regulatory compliance. They need integrated approaches centered on innovative methods and trends in design and manufacturing that help them focus on using environmentally friendly processes and creating green products. They need resources that help them participate in strategies for designing environmentally responsible products and methods, resources that provide a foundation for understanding and implementing principles of environmentally conscious engineering.

To help engineering practitioners and managers meet these needs, I envisioned a flexibly connected series of edited books, each devoted to a broad topic under the umbrella of Environmentally Conscious Engineering. The series started with three volumes that are closely linked—environmentally conscious mechanical design, environmentally conscious manufacturing, and environmentally conscious materials and chemicals processing. The series continues with this fourth volume, **Environmentally Conscious Alternative Energy Production**, and thereby turns toward a subject area more commonly associated among the general public with the future of the earth's climate and ramifications of climate changes while, of course, being of intense interest to a wide variety of engineers, scientists, and public policy makers. The topic carries additional weight because of the supply of fossil fuels, which generate the bulk of the world's power needs, is limited (although there is not consensus about the extent of the future supply), because major petroleum reserves are located in countries where there is political instability or the threat of it, and where, therefore, industrial nations believe they must retain a military presence to guarantee the future availability of oil to their economies. (The series will continue with a fifth volume on **Environmentally Conscious Transportation**, a sixth on **Environmentally Conscious Materials Handling**, plus a seventh on **Environmentally Conscious Fossil Energy Production**. The fourth through seventh volumes will be loosely linked, much like the first three design–manufacturing–materials volumes are. For example, a chapter on alternative fuels will appear in the transportation volumes, although it could fit quite well in the alternative energy volume.)

While many of the chapters in the books in the series are accessible to lay readers, the primary intended audience is practicing engineers and upper-level students in a number of areas—mechanical, chemical, industrial, manufacturing, plant, electrical, and environmental—as well as engineering managers. This audience is broad and multidisciplinary. In the case of power generation, an electrical or environmental engineer may be concerned with improving the performance of a plant that uses a particular technology, or an industrial or plant engineer may be involved in selecting a power generating technology for a new facility, and these practitioners be found in a wide a variety of organizations, including commercial facilities, institutions of higher learning, and consulting firms, as well as federal, state and local government agencies. A volume that covers a broad range of technologies is useful because every practitioner, researcher, and bureaucrat can't be an expert on every topic and may need to read an authoritative summary on a professional level of a subject that he or she is not intimately familiar with but may need to know about for a number of different reasons.

The Wiley Series in Environmentally Conscious Engineering is comprised of practical references for engineers who are seeking to answer a question, solve a problem, reduce a cost, or improve a system or facility. These books are not a research monographs. The purpose is to show readers what options are available in a particular situation and which option they might choose to solve problems at hand. I want these books to serve as a source of practical advice to readers. I would like them to be the first information resource a practicing engineer reaches for when faced with a new problem or opportunity—a place to turn to even before turning to other print sources, even any officially sanctioned ones, or to sites on the Internet. So the books have to be more than references or collections of background readings. In each chapter, readers should feel that they are in the hands of an experienced consultant who is providing sensible advice that can lead to beneficial action and results.

This fourth volume in the series, **Environmentally Conscious Alternative Energy Production**, offers technical descriptions of a number of different technologies so that readers may be able to not only evaluate them on their own merits, but also compare and contrast them, and, ultimately, choose from among them for a particular purpose. After an opening chapter that compares power generation technologies on an economic basis, the book presents chapters on the technologies, including solar, fuel cells, geothermal, wind, cogeneration, hydrogen, and coal, and closes with a chapter on using waste heat from power plants. Some experts may descry the lack of a chapter on nuclear power, but I excluded this technology because of uncertainty about environmentally friendly and politically palatable schemes for disposing of spent fuel rods, as well as the potential for mischief in diverting nuclear fuel to weaponry.

I asked the contributors, all of whom are located in North America, to provide short statements about the contents of their chapters and why the chapters are important. Here are their responses:

Todd Nemec (GE Energy, Schenectady, NY), who contributed the opening chapter on **Economic Comparisons of Power Generation Technologies**, writes, "this chapter discusses the components and applicability of Cost of Electricity models in addition to economic aspects of emissions regulation, nondispatchable (intermittent) generation, and cogeneration. From technology development to product design, applications/siting optimization, and operations, economic models are integral to environmentally friendly power-generation growth—as the basis for good decision making and increased customer value. Many environmentally friendly technologies have inherently low power density, affecting cost competitiveness, siting, and fuel availability/market viability concerns that aren't as significant in high power density thermal powerplants. On the opposite side, however, emissions control mechanisms such as cap and trade are efficient at delivering emissions control technologies to thermal plants as well as unlocking additional environmental value for emerging renewable platforms. High fidelity economic models and their effective use through technology selection, design, and applications will give newer, cleaner technologies the greatest chance to succeed.

The chapter on **Solar Energy Applications** by Jan F. Kreider (University of Colorado in Boulder, Colorado) has appeared in all three editions of the *Mechanical Engineers' Handbook*, published by Wiley. He writes, "Solar energy represents the most basic of renewable energies with its source both permanent and continuous. With terrestrial levels sufficient to supply all of the earth's energy needs, it will be the ultimate energy source after the fossil fuel era ends on earth. This chapter describes the resource and several practical methods for producing useful energy—including thermal energy and electricity—with engineering details."

Matthew W. Mensch (The Pennsylvania State University in University Park, Pennsylvania), who contributed the chapter on **Fuel Cells** (this chapter also appears in the *Mechanical Engineers' Handbook*, Third Edition), writes, "In the coming decades, mounting pressure from environmental, security, and economic concerns will usher into the mainstream a new age of power generation from alternative sources, gradually usurping traditional sources of energy from non-renewable fossil based fuels. While the specific future outcomes of each particular possibility are impossible to predict, a global future including use of fuel cells in many applications is now all but assured. Fuel cells will almost certainly play a key role in the future energy grid, potentially ending the century long reign of the internal combustion engine in transportation applications, supplanting rechargeable batteries for many portable applications such as cell phones and laptop computers, and providing reliable electricity and heat for stationary applications. The science of fuel cells is both fascinating and highly multidisciplinary, involving nearly all fields of engineering. There are different types of fuel cell systems, which operate under a wide range of conditions with highly varied materials, myriad system configurations, and a host of technical and economic challenges. Each particular system has fundamental advantages and limitations, which must be addressed before ubiquitous implementation can be achieved. This chapter describes the basic operating principles of each of the major fuel cell systems being developed today, and addresses the fundamental advantages and challenges remaining to be overcome. I hope this introduction can serve as a valuable starting point for engineers and managers looking for a technical overview of the potential for fuel cells as serious power generation sources."

Peter Blair (The National Academy of Sciences in Washington, DC), who contributed the chapter on **Geothermal Resources and Technology: An Introduction** (this chapter also appears in the *Mechanical Engineers' Handbook*, Third Edition), writes, "Geothermal energy, or heat extracted from the earth's interior, is often included in the portfolio of renewable energy sources that are considered to be more benign environmentally than fossil and nuclear energy sources. Geothermal energy has been used for centuries for cooking and heating and since the early 1900s for producing electric power. In its most economically attractive form, in the geologically rare situation when a very hot geothermal heat source and a water aquifer coincide, the resulting dry steam can be used to run a turbine directly for electric power generation. More commonly, but still relatively unusual geologically, hot water can be drawn from a geologic formation and its heat extracted into a secondary working fluid to once again produce electric power or to provide process heat. When geothermal resources are accessible they can be very economical and environmentally attractive alternatives to conventional energy sources. This chapter surveys the types of geothermal resources present around the world and the range of energy conversion technologies that can be employed in direct use of geothermal heat, in electric power generation, and by geothermal heat pumps for utilizing low-grade geothermal heat."

Todd Nemec (GE Power Systems in Schenectady, New York), who contributed the chapter, **Wind Turbines**, writes, "Wind turbine design carries many of the fundamental complexities of designing aircraft for airline service. Like aircraft, a balanced and integrated wind turbine design requires significant understanding of markets, aerodynamics/aeroelasticity, extreme and fatigue loading, controls, weight, noise, assembly/inbound transportation, and economic efficiency. This chapter introduces readers to wind energy's recent market growth, first-principles energy formulas, and conceptual design tradeoffs. The turbine power curve and siting discussions are a starting point for effectively matching turbine and site selection. Much of wind energy's improved economics are due to advancements in system-level design, component technology, and applications understanding. Market factors, incentives, environmental regulation, along with power industry contributions—such as increased energy storage, greater thermal powerplant flexibility, growth of distributed grid systems, and improved transmission infrastructure will also enable wind energy to reach higher levels of market entitlement.

Jerald Caton (Texas A&M University in College Station, Texas), who contributed the chapter on **Cogeneration**, writes, "Cogeneration is a technology that maximizes the utilization of the available energy from the combustion of fuels. A cogeneration system produces electrical power as well as thermal energy such as heat or cooling. The major motivations for considering cogeneration systems are monetary savings, energy savings, and the potential for lower emissions. Many facilities that have a need for electrical power and thermal energy are candidates for cogeneration. The technology for cogeneration systems is available and the concept is well developed. This chapter includes detailed discussions of the overall concept, descriptions of possible systems, a summary of relevant regulations, descriptions of economic evaluations, and comments on ownership and financing."

Elias K. Stefanokos (University of South Florida in Tampa, Florida), who contributed the chapter on **Hydrogen Energy** with Yogi Goswami, S. S. Srinivasan, and John T.Wolan, writes, "Fossil fuels are not renewable, they are limited in supply, their economic cost is continuously increasing, and their use is growing exponentially. Moreover, combustion of fossil fuels is causing global climate change and harming the environment in other ways as well, which points to the urgency of developing environmentally clean alternatives. Hydrogen is a good alternative to fossil fuels for the production, distribution and storage of energy. Automobiles can run on either hydrogen used as fuel in internal combustion engines or in fuel-cell cars or in hybrid configurations. Hydrogen is not an energy source but an energy carrier that holds tremendous potential to use renewable and clean energy options. It is not available in free form and must be dissociated from other molecules containing hydrogen such as natural gas or water. Once produced in free form it must be stored in a compressed or liquefied form, or in solid state materials. It is the purpose of this chapter to bring readers up to date on the state of the art and the obstacles that must be overcome to achieve cost effective production, storage and conversion of hydrogen."

James Butler (Dalhousie University in Halifax, Nova Scotia), who contributed the chapter on **Clean Power Generation from Coal** with Prabir Basu, writes, "Coal accounts for roughly 40% of the world's total electricity generating capacity and shows no signs of decreasing as emerging economies such as China and India, are fueling their rapid economic expansion with coal. With increased concern over global warming caused carbon dioxide and other harmful emissions of sulphur dioxide, nitrogen oxides and mercury from coal, there is a great deal of research and development taking place into new technologies that reduce the environmental impact of electricity generation from coal."

Herbert A. Hingley, III (University of Florida in Gainesville, Florida), who contributed the chapter on **Using Waste Heat from Power Plants**, writes, "This chapter discusses several examples of integrating power production with the utilization of the associated waste heat to accomplish some other function, such as space heating, domestic water heating, cooling, steam production or process heating. In addition to several domestic applications of combined heat and power, two new innovative systems for water purification and improved thermoelectric power production are reviewed. This chapter should be of importance to engineers and policy makers seeking innovative methods to better utilize our energy resources."

That ends the contributors' comments. I would like to express my heartfelt thanks to all of them for having taken the opportunity to work on this book. Their lives are terribly busy, and it is wonderful that they found the time to write thoughtful and complex chapters. I developed the book because I believed it could have a meaningful impact on the way many engineers approach their daily work, and I am gratified that the contributors thought enough of the idea that they were willing to participate in the project. Thanks also to my editor, Bob Argentieri, for his faith in the project from the outset. And a special note of thanks to my wife Arlene, whose constant support keeps me going.

Myer Kutz
Delmar, NY

**Environmentally Conscious
Alternative Energy Production**

CHAPTER 1

ECONOMIC COMPARISONS OF POWER GENERATION TECHNOLOGIES

Todd S. Nemec
GE Energy
Schenectady, New York

1 INTRODUCTION 1

2 MARKET GROWTH AND EMISSIONS 2

3 ECONOMIC EVALUATION 4

4 INTEGRATION OF INTERMITTENT RENEWABLES 7

5 COGENERATION 8

1 INTRODUCTION

Like power generation engineering calculations, power generation economic comparisons rely heavily on mathematical models. When only one or two major plant characteristics are being compared, a simple two-or three-term equation may be enough to reasonably predict their economic differences. Comparing all plant characteristics effectively across multiple generation platforms, and within different power markets/utility networks, may require thousands of inputs using high-frequency data.

Real energy environments include diversity and interactions of many variables, including sources of revenue, fuel cost and availability, emissions requirements, economic incentives, risk, and both initial and recurring costs—including logistics, labor rates, and investor return, among others. System designers and purchasers have different data requirements with respect to building appropriate design or applications economic models. Most system designers, however, begin their calculations by looking at variables from the owners' perspective—either by comparing revenue-requirements for a given profit, or by using market revenue rates to calculate profit.

This chapter will also look at calculations from an owners' outlook—using both revenue requirements also called *levelized cost of electricity* (COE), and projected power sale prices—when calculating cogeneration economics.

1

2   MARKET GROWTH AND EMISSIONS

Worldwide, electricity demand is projected to grow at 2.7 percent per year between 2003 and 2030,[1] India and China are at the high end of projections, 4.6 percent and 4.8 percent annual growth rates, respectively, while Japan's demand is projected to grow the slowest, at 0.7 percent.[1] Share projections for the types of fuels are listed in Figure 1.

Within the thermal power-plant sector, emissions have improved significantly during the last few decades due to both increased technology and tighter regulations. Market-driven regulations, such as cap and trade systems, are a cost-effective means to reduce emissions on a total system basis, and are often used in conjunction with individual power-plant limits. As an example, the 1990 U.S. Clean Air Act's Acid Rain Program uses cap and trade to reduce 2010 sulfur dioxide (SO_2, a precursor to acid rain) emissions from electric plants to 50 percent of 1980 levels.[2] Emissions reductions are achieved through overall cap levels—limits set by a central authority to meet health and/or environmental standards—broken into smaller, tradable, allowances. Companies or governments that produce above their allowance must buy credits to offset their emissions, while those that produce less than their allowance can sell them as credits.

Characteristics of several U.S. emissions programs are as follows:

- *U.S. Acid Rain NO_x Reduction Program.* Emissions are not capped, nor are trade allowances like the SO_2 program, but still overall goals are set by specifying maximum NO_x output levels relative to fuel energy input, based on boiler technology.[3]

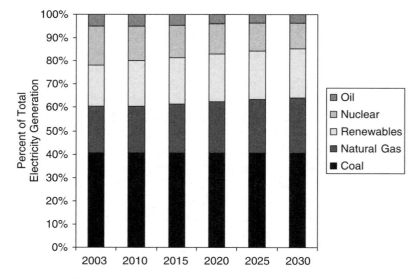

Figure 1   World electricity share projections by fuel source. (From Ref. 1.)

- *NO_x Budget Trading Program (NBP).* This is a cap and trade program involving eastern U.S. states, reducing NO_x during the summer ground-level ozone (smog) season. It is used to help States meet their EPA NOx State Implementation Plan (SIP) call.[3]
- *Clean Air Interstate Rule.* In 2005 the EPA established new or increased SIP requirements for 28 "upwind" states to reduce SO_2 and/or NO_x emissions.[4]
- *Clean Air Mercury Rule.* This cap and trade system was signed in 2005 and based on the Acid Rain Program. It will reduce 2018 coal-fired steam generating unit mercury emissions by approximately 70 percent over 1999 levels. Phase 1 of the plan, through 2017, takes advantage of mercury reduction through SO_2 and NO_x reductions in the Clean Air Interstate Rule.[5]

Although cap and trade policies promote innovative and cost-effective solutions, the increased compliance will eventually impact economics through one or more of the following factors: reduced net plant performance, increased capital cost, and/or increased operating cost. Within the Clean Air Act, economic choices for pulverized coal operators may include switching to a lower-sulfur coal and/or investing in more capable clean-up and control systems. Estimates for a typical flue gas desulfurization system, for example, which reduces SO_2 emissions of coal plants, are 0.14 cents/kWh operations cost and \$144/kW in installed capital cost.[6] Gas turbine nitric oxide emissions reductions are achieved through one or more precombustion, combustion, and postcombustion technologies. As technology improves, however, peak cycle temperatures are raised to increase thermal efficiency (reducing fuel burn and CO_2), which is in direct conflict with achieving lower NO_x. Higher thermal efficiency, combustor NO_x, cost of increased technology, and pre/post combustion treatment are all evaluated during design to find the lowest lifecycle cost while meeting emissions requirements.

Where traditional thermal generation is faced with capital and operating cost challenges for emissions avoidance and/or clean-up, nonthermal generation usually has a much better emissions entitlement, but must overcome the economics of a fundamentally lower power density (power per unit weight, or airflow). High pressures and temperatures in steam and gas turbine cycles enable high power density. Renewable generators such as wind and solar-photovoltaic have limited options to increase pressure and temperature, and must increase blade length, collector surface area, and/or efficiency to increase power. Low power density generally translates into greater land use/siting challenges, higher operations costs, and higher transportation costs per kWh. On the positive side, a failure or outage involving one wind turbine or solar panel within a farm means only a small-reduction total system output, allowing high overall system reliability.

Environmentally friendly sources of power will continue to be influenced by the following factors:

- Technology improvements
- Energy independence/security/diversity

- Increased scarcity and cost of alternatives
- Health and environmental/climate change:
  - Air and water
  - Global warming/greenhouse gases
- Creation of local manufacturing, construction, and maintenance jobs
- Improved transmission infrastructure to areas with significant renewable resources
- Regulatory standards and incentives, many driven by the factors already listed

Technology improvements are typically measured in their ability to generate energy at a low COE. For power generation equipment, this usually means lower cost, or higher kilowatts, to produce a lower $/kW, through greater thermal conversion efficiency, higher availability, lower operation, and maintenance costs, reduced emissions, or increased flexibility. One figure of merit that can account for these factors is cost of electricity.

3   ECONOMIC EVALUATION

Levelized COE is a useful single metric used to compare owners' life-cycle costs. COE converts all costs into a single cost of electricity rate, usually expressed in cents per kilowatt-hour. It is considered levelized because it reflects an equivalent single value, rather than first- or final-year rates of revenue. The resulting levelized cents per kWh is then already formatted to provide differences in net present value when given annual energy production, without any further corrections.

Fixed costs, or one-time capital costs, are converted to a rate—capital recovery—by multiplying by fixed charge rate (FCR) and dividing by annual kilowatt-hours. Variable costs are converted to a levelized annual equivalent and divided by annual kilowatt-hours. Levelized annual costs are derived either on a part-by-part basis, or for simplification in these examples: converted from annualized first-year costs, multiplying them with a levelizing factor.[7]

$$\text{Levelizing Factor} = CRF \times [1 - ((1+u)/(1+r))^n]/(r-u) \qquad (1)$$

where

CRF = Capital recovery factor, fraction
u = inflation rate, percent per year
r = discount rate, percent per year
n = term, years

Fixed charge rate is derived from a representative pro forma model that includes the cost of debt, equity, depreciation, escalation, tax rate, and other real project factors. Since the pro forma will already include escalation, capital recovery is already levelized. FCR is analogous to the equal-payment series capital recovery factor derived in economics and finance textbooks, which are a function of term, present value, and interest rate.

Capital recovery = FCR × Total capital requirement/Annual kilowatt-hours

Incremental fuel = Heat rate × Kilowatts × HHV/LHV × Fuel price

Figure 2 shows representative COE calculations for common thermal, nuclear, wind, and solar-photovoltaic generators. Capacity factor is defined as the portion of the year the unit is operating at its maximum power, and is shown at representative levels. The COE values reflect the revenue required to pay for both the fixed and variable costs of the plant. Note the differences in contribution of fuel, capital, and operations and maintenance to COE. The relative weighing of the individual components reflect both the risks and opportunities associated with each factor. Simple cycle gas turbines and combined-cycles are sensitive to fuel (fuel price and thermal efficiency), while solar PV, wind, and nuclear are highly sensitive to capital cost and factors included in FCR. The FCR used here is a constant value, which assumes that investor expected return, depreciation, risk, and other economic factors are considered equal on a net basis. This is acceptable for technology screening, but is not necessarily applicable in real-world analysis. Calculations have also been made on a *direct-unit basis*, assuming each alternative has a similar impact on the rest of the system. Higher-fidelity grid and power plant models would be required to understand system interactions.

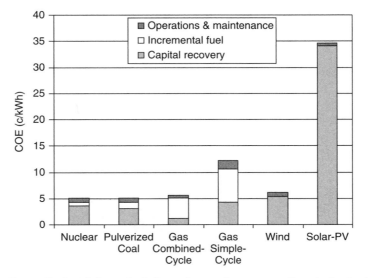

Figure 2 Cost of electricity calculations for various generating technologies. (From Ref. 12.)

It might seem intuitive that lowest total COE would produce the most cost-effective plant. The nature of varying day versus night and seasonal loads and plant output profiles, however, is best optimized (least cost) with a portfolio of diverse power plants designed to perform within a given range of low to high annual hours of operation. A load-duration curve (Figure 3) shows the annual hours spent at each load (shown as a percent of peak load) demanded for a representative transmission region. Assuming one generator was designed to serve this load, economics would be unfavorable, since capital recovery would be penalized for a high number of hours spent at part load, cycling to meet demand. Serving this load with multiple segments, traditionally shown as base, cyclic, and peak load, however, allows generators to optimize the characteristics of capital, fuel, and O&M costs to meet the load growth needs within different generation systems. Higher growth in any region of the curve creates higher demand for assets that can most economically meet the new operating profile.

Figure 4 shows COE versus hours for three representative thermal plants: simple cycle gas turbine, combined-cycle gas turbine, and pulverized coal. Because coal has the lowest COE at high hours of operation, coal would be the lowest cost-base load plant given this set of assumptions. Similarly, combined-cycle gas turbines are the lowest total cost platform for mid-range cyclic duty, and simple-cycle gas turbines will be selected for low hours, or peaking duty. As each platform is affected by increased technology, environmental regulation, fuel prices and availability, labor and materials costs, the placement of these curves, and relative competitiveness, will shift.

Within market-based grid regions, a common publication used to describe the cost of power is the price duration curve, Figure 5. This is an extension of Figures 3 and 4, and constructed similarly to the load duration curve, but it uses

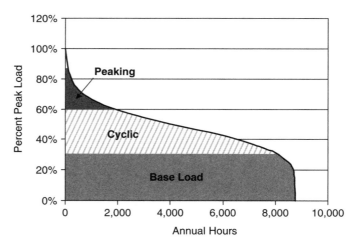

Figure 3 Load duration curve.

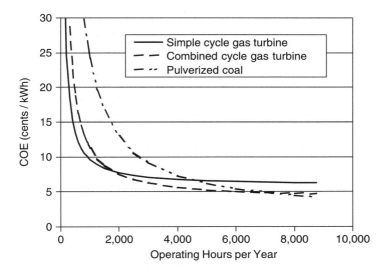

Figure 4 Cost of electricity versus annual hours of operation.

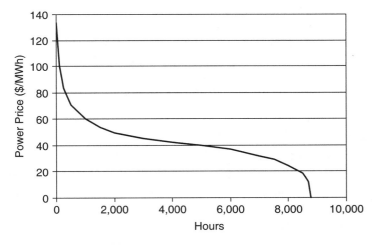

Figure 5 Price duration curve.

wholesale price instead of load. Using the shape of this curve, rational wholesale prices can be predicted based on variable operating cost.

4 INTEGRATION OF INTERMITTENT RENEWABLES

Power plants are typically dispatched in order of lowest to highest variable cost, either within a regulated environment or in a competitive marketplace, where bidding does not usually fall below variable cost. Renewables such as wind and solar are considered nondispatchable generators, meaning they—due to their intermittent fuel resource—are not able to provide power on demand. They do not usually have to compete with other generators for dispatch priority since their variable costs are lowest—because fuel is free. Intermittent renewable generators create unique system and siting issues within the rest of the grid. Concerns include transmission thermal, voltage, and stability limits/maximum intermittent capacity, necessary reserve capacity and allowable capacity payments, and the impact to nearby dispatchable generators.

Areas with good nondispatchable fuel resources are expected to have a higher upperlimit on their electricity potential. Western Denmark has over 20 percent of its electricity supplied by wind, part of this is due to good interconnection with a diverse mix of generating assets through the Nord Pool, which includes access to Norwegian Hydro resources.[8] Dispatchable generators may lose valuable hours of operation if wind is placed in its region, and may be expected to provide greater loading and unloading ramp rate to mitigate intermittent supply. Solar is regarded as generally more favorable than wind when comparing the coincidence of its peak output with load demand. Onshore wind in the United States is characterized as somewhat out of phase with peak demand, while offshore winds are much more coincident with demand.[9] Access to adequate transmission capacity, energy storage, highly flexible dispatchable power, demand-side management, increased distributed generation, wind farm curtailment, and improved wind forecasting will all help support higher levels of intermittent generation.[8] Growth in emerging forms of energy storage, such as hydrogen, will also help extend the market potential of nondispatchable generation.

5 COGENERATION

Combined heat and power production, when matched to the right set of thermal and energy demands, can provide both environmental and economic benefits over separate facilities. Power can be generated from both topping and bottoming cycles, with heat recovered below the topping cycle or used above a low-temperature power system. The following section describes characteristics of steam turbine and gas turbine plants, both of which provide topping cycle power.

In cogeneration applications, subcritical pulverized coal plants can deliver about 85 percent of fuel energy to heat and power, while gas turbines are capable of approximately 75 percent.[10,11] This compares to about 38 percent HHV and 59 percent LHV (combined-cycle) net electrical conversion rates, with no heat recovery, for the same two technologies, respectively. Cogeneration systems recover heat normally lost in the condenser of an electrical plant, increasing the overall system efficiency.

Two metrics have been defined to simplify comparisons between heat-only and cogeneration systems. The first, net heat to process, measures the net energy supplied to the process, and must be kept constant for a fixed plant design. Ideally, it should be varied during process plant conceptual design to include various generators, among other factors, in the total plant optimization. The second, fuel chargeable to power, measures the incremental heat rate of the net power relative to a heat-only system. This is directly comparable to the heat rate of a power-only system. Although low values of heat rate are usually preferred, other variables such as plant flexibility and fit, cost per kilowatt, operation and maintenance costs, fuel cost, and power price may help to define a better overall economic solution.

$$\text{Net heat to process (NHP)} = \text{Steam flow to process} \times \text{Enthalpy}$$
$$- \text{Process return flow} \times \text{Enthalpy}$$
$$- \text{Make up water} \times \text{Enthalpy} \qquad (2)$$

$$\text{Fuel chargeable to power (FCP)} = [\text{NHP/Packaged boiler efficiency}$$
$$- \text{Cogen fuel consumption}] \times \text{Net output}$$
$$(3)$$

Cogeneration fit can be described in terms of both their design and off-design envelopes defining output versus net heat to process. A sample design envelope is shown in Figure 6, including descriptions of each of the state points.

Several process variables influencing fit include the cyclic nature of steam and power demand: Many industrial processes may demand near-constant steam

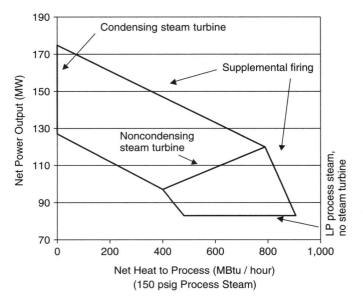

Figure 6 Design performance envelope for a gas turbine in cogeneration. (From Ref. 10.)

conditions, while commercial and educational users may have a more cyclic demand for heat and power. These can be mitigated through energy storage, import/export of additional power or steam, and/or greater attention to design integration with the cogeneration system and its features. On the generation side, gas turbines provide greater power output relative to steam production when compared to steam-turbine generators (greater than 85 kW/MBtu-hr for gas turbines).[10] They also have reduced maximum steam capability on a hot-day relative to ISO. Inlet devices, such as evaporative coolers or chillers, and supplemental firing could be used to help restore hot day steam production through higher exhaust flow and temperature, respectively. Table 1 lists a sample

Table 1 Economic Comparisons of Cogeneration Alternatives

Alternative	Units	Packaged Boiler	GT	GT + ST	GT + ST
Number of GTs		NA	1	1	2
HRSG pressure levels		NA	1	2	2
Steam turbine		NA	None	Noncondensing	Extraction/Condensing
Net fuel	Mbtu/h HHV	452	508	508	1494
	MkJ/h HHV	477.4	536	536	1576
Net power	MW	NA	84.4	98.6	228.5
Heat rate	BTU/kWh LHV	NA	10430	8928	7705
Estimated installed cost	$MM	14	39	60	124
Steam to process					
Pressure	psig	150	150	150	150
Temperature	deg F	365	365	365	365
Flow	lb/hr	373,000	373,000	373,000	373,000
Net heat to process	MMBtu/h	385	385	385	385
Fuel chargeable to power (FCP)	BTU/kWh HHV	NA	6206	5312	6565
Incremental cost of fuel	cents/kWh	NA	3.72	3.19	3.94
Cost of purchased electricity	cents/kWh		5.00	5.00	5.00
Incremental cost of maintenance	cents/kWh		0.17	0.15	0.15
Installed incremental capital cost	$/kW		296.2	466.5	481.4
Incremental cost of electricity (fuel and maintenance)	cents/kWh		3.89	3.34	4.09
Savings per kW of Cogen System (cost of purchased − incremental cost of electricity)			1.11	1.66	0.91
$/kW Annually			83	125	68
$/year			7,003,000	12,294,000	15,615,000
Simple payback (years)		Base	3.57	3.74	7.04
Present value		Base	34,620,000	58,666,000	22,939,000
Internal rate of return before taxes, depreciation		Base	28%	26%	13%

Source: Based on calculations in Ref. 10.

comparison between installing a packaged boiler versus three variants of a GE MS7001EA (gas-turbine) based cogeneration cycle. Results assume the first-year fuel price and power price are $6.00/MMBtu and 6.25 cents/kWh, respectively, and 7,500 hours/year utilization. Results do not reflect the cost or performance at any particular site, but are designed to show the methods and factors included in preliminary economic screening calculations.

REFERENCES

1. U.S. DOE Energy Information Association (EIA), "International Energy Outlook 2006," Washington, D.C., 2006.

2. U.S. EPA, EPA-430F-02-009, "Clearing the Air—The Facts About Capping and Trading Emissions," Washington D.C., March 2002.

3. U.S. EPA, EPA-430-R-04-010, "NO_x Budget Trading Program, 2003 Progress and Compliance Report," Washington D.C., August 2004.

4. U.S. EPA, "Rule to Reduce Interstate Transport of Fine Particulate Matter and Ozone, Revisions to Acid Rain Program, Revisions to NO_x SIP Call," *Federal Register*, **70**(91) (May 12, 2005).

5. U.S. EPA, Rule T6560-50-P, "Standards of Performance for New and Existing Stationary Sources: Electric Utility Steam Generating Units," Washington D.C.

6. U.S. DOE/EIA-0554 (2006), "Assumptions to the Annual Energy Outlook," Washington D.C., March 2006.

7. W. D. Marsh, *Economics of Electric Utility Power Generation*, Oxford University Press, Oxford, 1980.

8. "Variability of Wind Power and Other Renewables," International Energy Agency, Paris, June 2005.

9. "The Effects of Integrating Wind Power on Transmission System Planning, Reliability, and Operations," prepared for the New York State Energy Research and Development Authority, GE Energy Consulting, 2005.

10. Bob Fisk and Robert VanHousen, GER3430f, "Cogeneration Application Considerations," GE report.

11. Harry G. Stoll, *Least-Cost Electric Utility Planning*, John Wiley & Sons, New York, 1989.

12. EIA, "Annual Energy Outlook, 2006," U.S. DOE Energy Information Association, 2006.

CHAPTER 2

SOLAR ENERGY APPLICATIONS

Jan F. Kreider
Kreider and Associates, LLC
and
Joint Center for Energy Management
University of Colorado
Boulder, Colorado

1	**SOLAR ENERGY**			3.2	Mechanical Solar Space Heating
	AVAILABILITY	**13**			Systems 42
	1.1 Solar Geometry	13		3.3	Passive Solar Space Heating
	1.2 Sunrise and Sunset	17			Systems 43
	1.3 Solar Incidence Angle	19		3.4	Solar Ponds 44
	1.4 Quantitative Solar Flux			3.5	Solar Thermal Power
	Availability	22			Production 47
2	**SOLAR THERMAL**			3.6	Other Thermal Applications 48
	COLLECTORS	**28**		3.7	Performance Prediction for Solar
	2.1 Flat-Plate Collectors	28			Thermal Processes 48
	2.2 Concentrating Collectors	33	**4**	**PHOTOVOLTAIC SOLAR**	
	2.3 Collector Testing	37		**ENERGY APPLICATIONS**	**49**
3	**SOLAR THERMAL**				
	APPLICATIONS	**39**			
	3.1 Solar Water Heating	39			

1 SOLAR ENERGY AVAILABILITY

Solar energy is defined as that radiant energy transmitted by the sun and intercepted by Earth. It is transmitted through space to Earth by electromagnetic radiation with wavelengths ranging between 0.20 and 15 μm. The availability of solar flux for terrestrial applications varies with season, time of day, location, and collecting surface orientation. In this chapter we shall treat these matters analytically.

1.1 Solar Geometry

Two motions of the Earth relative to the sun are important in determining the intensity of solar flux at any time—Earth's rotation about its axis and the annual motion of Earth and its axis about the sun. Earth rotates about its axis once

13

each day. A solar day is defined as the time that elapses between two successive crossings of the local meridian by the sun. The local meridian at any point is the plane formed by projecting a north–south longitude line through the point out into space from the center of the earth. The length of a solar day on the average is slightly less than 24 hours, owing to the forward motion of Earth in its solar orbit. Any given day will also differ from the average day owing to orbital eccentricity, axis precession, and other secondary effects embodied in the equation of time described below.

Declination and Hour Angle

The Earth's orbit about the sun is elliptical with eccentricity of 0.0167. This results in variation of solar flux on the outer atmosphere of about 7 percent over the course of a year. Of more importance is the variation of solar intensity

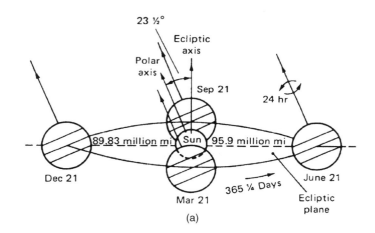

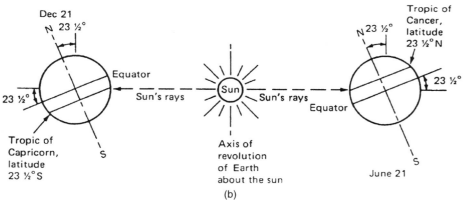

Figure 1   (a) Motion of Earth about the sun. (b) Location of tropics. Note that the sun is so far from Earth all the rays of the sun may be considered as parallel to one another when they reach Earth.

caused by the inclination of Earth's axis relative to the ecliptic plane of its orbit. The angle between the ecliptic plane and the Earth's equatorial plane is 23.45°. Figure 1 shows this inclination schematically.

Earth's motion is quantified by two angles varying with season and time of day. The angle varying on a seasonal basis that is used to characterize Earth's location in its orbit is called the solar *declination*. It is the angle between the Earth–sun line and the equatorial plane, as shown in Figure 2. The declination δ_s is taken to be positive when the Earth–sun line is north of the equator and negative otherwise. The declination varies between +23.45° on the summer solstice (June 21 or 22) and −23.45° on the winter solstice (December 21 or 22). The declination is given by

$$\sin \delta_s = 0.398 \cos[0.986(N - 173)] \tag{1}$$

in which N is the day number, counted from January 1.

The second angle used to locate the sun is the solar-hour angle. Its value is based on the nominal 360° rotation of Earth occurring in 24 hours. Therefore, 1 hour is equivalent to an angle of 15°. The hour angle is measured from zero at solar noon. It is denoted by h_s and is positive before solar noon and negative after noon in accordance with the right-hand rule. For example 2:00 PM corresponds to $h_s = -30°$ and 7:00 AM corresponds to $h_s = +75°$.

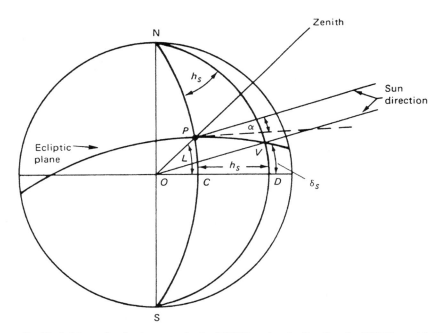

Figure 2 Definition of solar-hour angle h_s (*CND*), solar declination δ_s (*VOD*), and latitude L (*POC*): *P*, site of interest. (Modified from J. F. Kreider and F. Kreith. *Solar Heating and Cooling*, revised 1st ed., Hemisphere, Washington, DC, 1977.)

Solar time, as determined by the position of the sun, and clock time differ for two reasons. First, the length of a day varies because of the ellipticity of Earth's orbit; and second, standard time is determined by the standard meridian passing through the approximate center of each time zone. Any position away from the standard meridian has a difference between solar and clock time given by [(local longitude − standard meridian longitude)/15] in units of hours. Therefore, solar time and local standard time (LST) are related by

$$\text{Solar time} = \text{LST} - \text{EoT} - (\text{Local longitude} - \text{Standard meridian longitude})/15 \tag{2}$$

in units of hours. EoT is the equation of time which accounts for difference in day length through a year and is given by

$$\text{EoT} = 12 + 0.1236 \sin x - 0.0043 \cos x + 0.1538 \sin 2x + 0.0608 \cos 2x \tag{3}$$

in units of hours. The parameter x is

$$x = \frac{360(N - 1)}{365.24} \tag{4}$$

where N is the day number.

Solar Position

The sun is imagined to move on the celestial sphere, an imaginary surface centered at Earth's center and having a large but unspecified radius. Of course, it is Earth that moves, not the sun, but the analysis is simplified if one uses this Ptolemaic approach. No error is introduced by the moving sun assumption, since the relative motion is the only motion of interest. Since the sun moves on a spherical surface, two angles are sufficient to locate the sun at any instant. The two most commonly used angles are the solar-altitude and azimuth angles (see Figure 3) denoted by α and a_s, respectively. Occasionally, the solar-zenith angle, defined as the complement of the altitude angle, is used instead of the altitude angle.

The solar-altitude angle is related to the previously defined declination and hour angles by

$$\sin \alpha = \cos L \cos \delta_s \cos h_s + \sin L + \sin \delta_s \tag{5}$$

in which L is the latitude, taken positive for sites north of the equator and negative for sites south of the equator. The altitude angle is found by taking the inverse sine function of equation (5).

The solar-azimuth angle is given by

$$\sin a_s = \frac{\cos \delta_s \sin h_s}{\cos \alpha} \tag{6}$$

To find the value of a_s, the location of the sun relative to the east−west line through the site must be known. This is accounted for by the following two

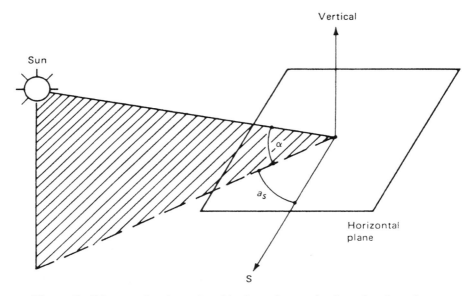

Figure 3 Diagram showing solar-altitude angle α and solar-azimuth angle a_s.

expressions for the azimuth angle:

$$a_s = \sin^{-1}\left(\frac{\cos \delta_s \sin h_s}{\cos \alpha}\right), \quad \cos h_S > \frac{\tan \delta_s}{\tan L} \tag{7}$$

$$a_s = 180° - \sin^{-1}\left(\frac{\cos \delta_s \sin h_s}{\cos \alpha}\right), \quad \cos h_s < \frac{\tan \delta_s}{\tan L} \tag{8}$$

Table 1 lists typical values of altitude and azimuth angles for latitude $L = 40°$. Complete tables are contained in Refs. 1 and 2.

1.2 Sunrise and Sunset

Sunrise and sunset occur when the altitude angle $\alpha = 0$. As indicated in Figure 4, this occurs when the center of the sun intersects the horizon plane. The hour angle for sunrise and sunset can be found from equation (5) by equating α to zero. If this is done, the hour angles for sunrise and sunset are found to be

$$h_{sr} = \cos^{-1}(-\tan L \tan \delta_s) = -h_{ss} \tag{9}$$

in which h_{sr} is the sunrise hour angle and h_{ss} is the sunset hour angle.

Figure 4 shows the path of the sun for the solstices and the equinoxes (length of day and night are both 12 hours on the equinoxes). This drawing indicates the very different azimuth and altitude angles that occur at different times of year at identical clock times. The sunrise and sunset hour angles can be read from the figures where the sun paths intersect the horizon plane.

Table 1   Solar Position for 40°N Latitude

Date	Solar Time AM	Solar Time PM	Altitude	Azimuth	Date	Solar Time AM	Solar Time PM	Altitude	Azimuth
January 21	8	4	8.1	55.3	July 21	5	7	2.3	115.2
	9	3	16.8	44.0		6	6	13.1	106.1
	10	2	23.8	30.9		7	5	24.3	97.2
	11	1	28.4	16.0		8	4	35.8	87.8
		12	30.0	0.0		9	3	47.2	76.7
February 21	7	5	4.8	72.7		10	2	57.9	61.7
	8	4	15.4	62.2		11	1	66.7	37.9
	9	3	25.0	50.2			12	70.6	0.0
	10	2	32.8	35.9	August 21	6	6	7.9	99.5
	11	1	38.1	18.9		7	5	19.3	90.9
		12	40.0	0.0		8	4	30.7	79.9
March 21	7	5	11.4	80.2		9	3	41.8	67.9
	8	4	22.5	69.6		10	2	51.7	52.1
	9	3	32.8	57.3		11	1	59.3	29.7
	10	2	41.6	41.9			12	62.3	0.0
	11	1	47.7	22.6	September 21	7	5	11.4	80.2
		12	50.0	0.0		8	4	22.5	69.6
April 21	6	6	7.4	98.9		9	3	32.8	57.3
	7	5	18.9	89.5		10	2	41.6	41.9
	8	4	30.3	79.3		11	1	47.7	22.6
	9	3	41.3	67.2			12	50.0	0.0
	10	2	51.2	51.4	October 21	7	5	4.5	72.3
	11	1	58.7	29.2		8	4	15.0	61.9
		12	61.6	0.0		9	3	24.5	49.8
May 21	5	7	1.9	114.7		10	2	32.4	35.6
	6	6	12.7	105.6		11	1	37.6	18.7
	7	5	24.0	96.6			12	39.5	0.0
	8	4	35.4	87.2	November 21	8	4	8.2	55.4
	9	3	46.8	76.0		9	3	17.0	44.1
	10	2	57.5	60.9		10	2	24.0	31.0
	11	1	66.2	37.1		11	1	28.6	16.1
		12	70.0	0.0			12	30.2	0.0
June 21	5	7	4.2	117.3	December 21	8	4	5.5	53.0
	6	6	14.8	108.4		9	3	14.0	41.9
	7	5	26.0	99.7		10	2	20.0	29.4
	8	4	37.4	90.7		11	1	25.0	15.2
	9	3	48.8	80.2			12	26.6	0.0
	10	2	59.8	65.8					
	11	1	69.2	41.9					
		12	73.5	0.0					

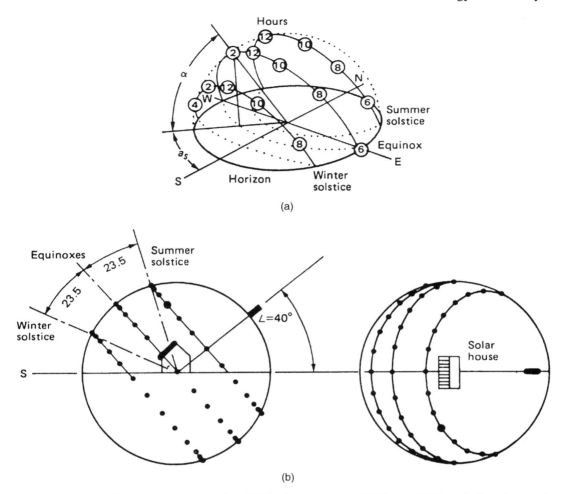

Figure 4 Sun paths for the summer solstice (6/21), the equinoxes (3/21 and 9/21), and the winter solstice (12/21) for a site at 40°N; (a) isometric view; (b) elevation and plan views.

1.3 Solar Incidence Angle

For a number of reasons, many solar collection surfaces do not directly face the sun continuously. The angle between the sun–Earth line and the normal to any surface is called the incidence angle.

The intensity of off-normal solar radiation is proportional to the cosine of the incidence angle. For example, Figure 5 shows a fixed planar surface with solar radiation intersecting the plane at the incidence angle i measured relative to the surface normal. The intensity of flux at the surface is $I_b \times \cos i$, where I_b is the beam radiation along the sun–Earth line; I_b is called the direct, normal radiation. For a fixed surface such as that in Figure 5 facing the equator, the incidence angle

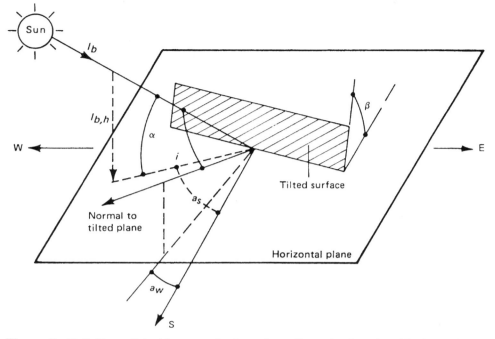

Figure 5   Definition of incidence angle i, surface tilt angle β, solar-altitude angle α, wall-azimuth angle a_w, and solar-azimuth angle a_s for a non–south-facing tilted surface. Also shown is the beam component of solar radiation I_b and the component of beam radiation $I_{b,h}$ on a horizontal plane.

is given by

$$
\begin{aligned}
\cos i = {} & \sin \delta_1 (\sin L \cos \beta - \cos L \sin \beta \cos a_w) \\
& + \cos \delta_s \cos h_s (\cos L \cos \beta + \sin L \sin \beta \cos a_w) \qquad (10) \\
& + \cos \delta_s \sin \beta \ \sin a_w \sin h_s
\end{aligned}
$$

in which a_w is the "wall" azimuth angle and β is the surface tilt angle relative to the horizontal plane, both as shown in Figure 5.

For fixed surfaces that face due south, the incidence angle expression simplifies to

$$
\cos i = \sin(L - \beta) \sin \delta_s + \cos(L - \beta) \cos \delta_s \cos h_s \qquad (11)
$$

A large class of solar collectors move in some fashion to track the sun's diurnal motion, thereby improving the capture of solar energy. This is accomplished by reduced incidence angles for properly tracking surfaces vis-á-vis a fixed surface for which large incidence angles occur in the early morning and late afternoon (for generally equator-facing surfaces). Table 2 lists incidence angle expressions for nine different types of tracking surfaces. The term *polar axis* in this table

Table 2 Solar Incidence Angle Equations for Tracking Collectors

Description	Axis (Axes)	Cosine of Incidence Angle ($\cos i$)
Movements in altitude and azimuth	Horizontal axis and vertical axis	1
Rotation about a polar axis and adjustment in declination	Polar axis and declination axis	1
Uniform rotation about a polar axis	Polar axis	$\cos \delta_s$
East–west horizontal	Horizontal, east–west axis	$\sqrt{1 - \cos^2 \alpha \sin^2 \alpha_s}$
North–south horizontal	Horizontal, north–south axis	$\sqrt{1 - \cos^2 \alpha \cos^2 \alpha_s}$
Rotation about a vertical axis of a surface tilted upward L (latitude) degrees	Vertical axis	$\sin(\alpha + L)$
Rotation of a horizontal collector about a vertical axis	Vertical axis	$\sin \alpha$
Rotation of a vertical surface about a vertical axis	Vertical axis	$\cos \alpha$
Fixed tubular collector	North–south tiled up at angle β	$\sqrt{1 - [\sin(\beta - L) \cos \delta_s \cos h_s + \cos(\beta - L) \sin \delta_s]^2}$

21

refers to an axis of rotation directed at the north or south pole. This axis of rotation is tilted up from the horizontal at an angle equal to the local latitude. It is seen that normal incidence can be achieved (i.e., cos i = 1) for any tracking scheme for which two axes of rotation are present. The polar case has relatively small incidence angles as well, limited by the declination to ±23.45°. The mean value of cos i for polar tracking is 0.95 over a year, nearly as good as the two-axis case for which the annual mean value is unity.

1.4   Quantitative Solar Flux Availability

The previous section has indicated how variations in solar flux produced by seasonal and diurnal effects can be quantified. However, the effect of weather on solar energy availability cannot be analyzed theoretically; it is necessary to rely on historical weather reports and empirical correlations for calculations of actual solar flux. In this section this subject is described along with the availability of solar energy at the edge of the atmosphere—a useful correlating parameter, as seen shortly.

Extraterrestrial Solar Flux

The flux intensity at the edge of the atmosphere can be calculated strictly from geometric considerations if the direct-normal intensity is known. Solar flux incident on a terrestrial surface, which has traveled from sun to earth with negligible change in direction, is called *beam radiation* and is denoted by $I_{b,0}$. The extraterrestrial value of I_b averaged over a year is called the *solar constant*, denoted by I_s. Its value is 429 Btu/hr·ft^2 or 1353 W/m^2. Owing to the eccentricity of Earth's orbit, however, the extraterrestrial beam radiation intensity varies from this mean solar constant value. The variation of $I_{b,0}$ over the year is given by

$$I_{b,0}(N) = \left[1 + 0.034 \cos\left(\frac{360N}{265}\right)\right] \times I_{sc} \tag{12}$$

in which N is the day number as before.

In subsequent sections the total daily, extraterrestrial flux will be particularly useful as a nondimensionalizing parameter for terrestrial solar flux data. The instantaneous solar flux on a horizontal, extraterrestrial surface is given by

$$I_{b,h\theta} = I_{b,0}(N) \sin\alpha \tag{13}$$

as shown in Figure 5. The daily total, horizontal radiation is denoted by I_0 and is given by

$$I_0(N) = \int_{I_w}^{I_s} I_{b,0}(N) \sin\alpha \, dt \tag{14}$$

$$I_0(N) = \frac{24}{\pi} I_{s0} \left[1 + 0.034 \cos\left(\frac{360N}{265}\right)\right]$$
$$\times (\cos L \cos\delta_s \sin h_{sr} + h_{sr} \sin L \sin\delta_s) \tag{15}$$

Table 3 Average Extraterrestrial Radiation on a Horizontal Surface $\overline{H}_0$ in SI Units and in English Units Based on a Solar Constant of 429 Btu/hr·ft² or 1.353 kW/m²

Latitude, Degrees	January	February	March	April	May	June	July	August	September	October	November	December
SI units, W·hr/m²·day												
20	7415	8397	9552	10,422	10,801	10,868	10,794	10,499	9791	8686	7598	7076
25	6656	7769	9153	10,312	10,936	11,119	10,988	10,484	9494	8129	6871	6284
30	5861	7087	8686	10,127	11,001	11,303	11,114	10,395	9125	7513	6103	5463
35	5039	6359	8153	9869	10,995	11,422	11,172	10,233	8687	6845	5304	4621
40	4200	5591	7559	9540	10,922	11,478	11,165	10,002	8184	6129	4483	3771
45	3355	4791	6909	9145	10,786	11,477	11,099	9705	7620	5373	3648	2925
50	2519	3967	6207	8686	10,594	11,430	10,981	9347	6998	4583	2815	2100
55	1711	3132	5460	8171	10,358	11,352	10,825	8935	6325	3770	1999	1320
60	963	2299	4673	7608	10,097	11,276	10,657	8480	5605	2942	1227	623
65	334	1491	3855	7008	9852	11,279	10,531	8001	4846	2116	544	97

Table 3 (*continued*)

English units, Btu/ft²·day

Latitude, Degrees	January	February	March	April	May	June	July	August	September	October	November	December
20	2346	2656	3021	3297	3417	3438	3414	3321	3097	2748	2404	2238
25	2105	2458	2896	3262	3460	3517	3476	3316	3003	2571	2173	1988
30	1854	2242	2748	3204	3480	3576	3516	3288	2887	2377	1931	1728
35	1594	2012	2579	3122	3478	3613	3534	3237	2748	2165	1678	1462
40	1329	1769	2391	3018	3455	3631	3532	3164	2589	1939	1418	1193
45	1061	1515	2185	2893	3412	3631	3511	3070	2410	1700	1154	925
50	797	1255	1963	2748	3351	3616	3474	2957	2214	1450	890	664
55	541	991	1727	2585	3277	3591	3424	2826	2001	1192	632	417
60	305	727	1478	2407	3194	3567	3371	2683	1773	931	388	197
65	106	472	1219	2217	3116	3568	3331	2531	1533	670	172	31

in which I_{s0} is the solar constant. The extraterrestrial flux varies with time of year via the variations of δ_s and h_{sr} with time of year. Table 3 lists the values of extraterrestrial, horizontal flux for various latitudes averaged over each month. The monthly averaged, horizontal, extraterrestrial solar flux is denoted by H_0.

Terrestrial Solar Flux

Values of instantaneous or average terrestrial solar flux cannot be predicted accurately owing to the complexity of atmospheric processes that alter solar flux magnitudes and directions relative to their extraterrestrial values. Air pollution, clouds of many types, precipitation, and humidity all affect the values of solar flux incident on Earth. Rather than attempting to predict solar availability accounting for these complex effects, one uses long-term historical records of terrestrial solar flux for design purposes.

The U.S. National Weather Service (NWS) records solar flux data at a network of stations in the United States. The pyranometer instrument, as shown in Figure 6, is used to measure the intensity of horizontal flux. Various data sets are available from the National Climatic Center (NCC) of the NWS. Prior to 1975, the solar network was not well maintained; therefore, the pre-1975 data were rehabilitated in the late 1970s and are now available from the NCC on magnetic media. Also, for the period 1950 to 1975, synthetic solar data have been generated for approximately 250 U.S. sites where solar flux data were not recorded. The predictive scheme used is based on other widely available meteorological data. Finally, from 1977 to the mid-1990s the NWS recorded hourly solar flux data at a 38-station network with improved instrument maintenance. In addition to horizontal flux, direct-normal data were recorded and archived at the NCC. Figure 7 is a contour map of annual, horizontal flux for the United States based on recent data.

The principal difficulty with using NWS solar data is that they are available for horizontal surfaces only. Solar-collecting surfaces normally face the general direction of the sun and are, therefore, rarely horizontal. It is necessary to convert measured horizontal radiation to radiation on arbitrarily oriented collection surfaces. This is done using empirical approaches to be described.

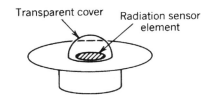

Figure 6 Schematic drawing of a pyranometer used for measuring the intensity of total (direct plus diffuse) solar radiation.

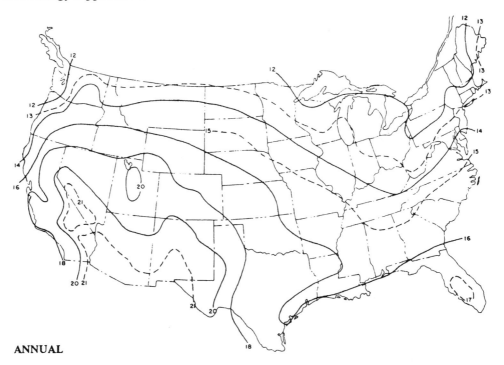

ANNUAL

*1 mJ/m² = 88.1 Btu/ft².

Figure 7 Mean daily solar radiation on a horizontal surface in megajoules per square meter for the continental United States.

Hourly Solar Flux Conversions

Measured, horizontal solar flux consists of both beam and diffuse radiation components. Diffuse radiation is that scattered by atmospheric processes; it intersects surfaces from the entire sky dome, not just from the direction of the sun. Separating the beam and diffuse components of measured, horizontal radiation is the key difficulty in using NWS measurements.

The recommended method for finding the beam component of total (i.e., beam plus diffuse) radiation is described in Ref. 1. It makes use of the parameter k_T called the clearness index and defined as the ratio of terrestrial to extraterrestrial hourly flux on a horizontal surface. In equation form k_T is

$$k_T \equiv \frac{I_h}{I_{b,h0}} = \frac{I_b}{I_{b,0}(N)\sin\alpha} \tag{16}$$

in which I_h is the measured, total horizontal flux. The beam component of the terrestrial flux is then given by the empirical equation

$$I_b = (ak_r + b)I_{b,0}(N) \tag{17}$$

Table 4 Empirical Coefficients for Equation (17)

Interval for K_T	a	b
0.00, 0.05	0.04	0.00
0.05, 0.15	0.01	0.002
0.15, 0.25	0.06	−0.006
0.25, 0.35	0.32	−0.071
0.35, 0.45	0.82	−0.246
0.45, 0.55	1.56	−0.579
0.55, 0.65	1.69	−0.651
0.65, 0.75	1.49	−0.521
0.75, 0.85	0.27	0.395

in which the empirical constants a and b are given in Table 4. Having found the beam radiation, the horizontal diffuse component $I_{d,h}$ is found by the simple difference

$$I_{d,h} = I_h - I_b \sin \alpha \qquad (18)$$

The separate values of horizontal beam and diffuse radiation can be used to find radiation on any surface by applying appropriate geometric *tilt factors* to each component and forming the sum accounting for any radiation reflected from the foreground. The beam radiation incident on any surface is simply $I_b \cos i$. If one assumes that the diffuse component is isotropically distributed over the sky dome, the amount intercepted by any surface tilted at an angle β is $I_{d,h} \cos^2(\beta/2)$. The total beam and diffuse radiation intercepted by a surface I_c is then

$$I_c = I_b \cos i + I_{d,h} \cos^2(\beta/2) + \rho I_h \sin^2(\beta/2) \qquad (19)$$

The third term in this expression accounts for flux reflected from the foreground with reflectance ρ.

Monthly Averaged, Daily Solar Flux Conversions

Most performance prediction methods make use of monthly averaged solar flux values. Horizontal flux data are readily available, but monthly values on arbitrarily positioned surfaces must be calculated using a method similar to that previously described for hourly tilted surface calculations. The monthly averaged flux on a tilted surface I_c is given by

$$\overline{I}_c = \overline{R}\overline{H}_h \qquad (20)$$

in which $\overline{H}_h$ is the monthly averaged, daily total of horizontal solar flux and R is the overall tilt factor given by equation (21) for a fixed, equator-facing surface:

$$\overline{R} = \left(1 - \frac{\overline{D}_h}{\overline{H}_h}\right)\overline{R}_h + \frac{\overline{D}_h}{\overline{H}_h}\cos^2 \frac{\beta}{2} + \rho \sin^2 \frac{\beta}{2} \qquad (21)$$

The ratio of monthly averaged diffuse, D, to total flux, $\overline{D}_h/\overline{H}_h$, is given by

$$\frac{\overline{D}_h}{\overline{H}_h} = 0.775 + 0.347\left(h_{sr} - \frac{\pi}{2}\right) - \left[0.505 + 0.261\left(h_{sr} - \frac{\pi}{2}\right)\right]$$
$$\times \cos\left[(\overline{K}_T - 0.9)\frac{360}{\pi}\right] \tag{22}$$

in which $\overline{K}_T$ is the monthly averaged clearness index analogous to the hourly clearness index. $\overline{K}_T$ is given by

$$\overline{K}_T \equiv \overline{H}_h/\overline{H}_0$$

where $\overline{H}_0$ is the monthly averaged, extraterrestrial radiation on a horizontal surface at the same latitude at which the terrestrial radiation $\overline{H}_h$ was recorded. The monthly averaged beam radiation tilt factor $\overline{R}_b$ is

$$\overline{R}_b = \frac{\cos(L - \beta)\cos\delta_s\sin h'_{sr} + h'_{sT}\sin(L - \beta)\sin\delta_s}{\cos L\cos\delta_s\sin h_{sr} + h_{sr}\sin L\sin\delta_s} \tag{23}$$

The sunrise hour angle is found from equation (9) and the value of h'_{sr} is the smaller of (1) the sunrise hour angle h_{sr} and (2) the collection surface sunrise hour angle found by setting $i = 90°$ in equation (11). That is, h'_{sr} is given by

$$h'_{sr} = \min\{\cos^{-1}[-\tan L\tan\delta_s], \cos^{-1}[-\tan(L - \beta)\tan\delta_s]\} \tag{24}$$

Expressions for solar flux on a tracking surface on a monthly averaged basis are of the form

$$\overline{I}_c = \left[r_T - r_d\left(\frac{\overline{D}_h}{\overline{H}_h}\right)\right]\overline{H}_h \tag{25}$$

in which the tilt factors r_T and r_d are given in Table 5.[2] Equation (22) is to be used for the diffuse to total flux ratio $\overline{D}_h/\overline{H}_h$.

2   SOLAR THERMAL COLLECTORS

The principal use of solar energy is in the production of heat at a wide range of temperatures matched to a specific task to be performed. The temperature at which heat can be produced from solar radiation is limited to about 6,000°F by thermodynamic, optical, and manufacturing constraints. Between temperatures near ambient and this upper limit very many thermal collector designs are employed to produce heat at a specified temperature. This section describes the common thermal collectors.

2.1   Flat-Plate Collectors

From a production volume standpoint, the majority of installed solar collectors are of the flate-plate design; these collectors are capable of producing heat at temperatures up to 100°C. Flat-plate collectors are so named since all components are planar. Figure 8a is a partial isometric sketch of a liquid-cooled

Table 5 Concentrator Tilt Factors (monthly averaged) (From Ref.~2.)

Collector Type	$r_T^{a,b,c,d}$	r_d^e
Fixed aperture concentrators that do not view the foreground	$[\cos(L-\beta)/(d\cos L)]\{-ah_{\text{coll}}\cos h_{sr}(i=90°)$ $+[a-b\cos h_{sr}(i=90°)]\sin h_{\text{coll}}$ $+(b/2)(\sin h_{\text{coll}}\cos h_{\text{coll}}+h_{\text{coll}})\}$	$(\sin h_{\text{coll}}/d)\{[\cos(L+\beta)/\cos L]-[1/(\text{CR})]\}+(h_{\text{coll}}/d)$ $\times\{[\cos h_{sr}/(\text{CR})]-[\cos(L-\beta)/\cos L]\cos h_{sr}(i=90°)\}$
East–west axis tracking[f]	$(1/d)\int_0^{h_{\text{coll}}}\{[(a+b\cos x)/\cos L]$ $\times\sqrt{\cos^2 x+\tan^2\delta_s}\}dx$	$(1/d)\int_0^{h_{\text{coll}}}\{[1/\cos L]\sqrt{\cos^2 x+\tan^2\delta_s}-[1/(\text{CR})]\}$ $\times[\cos x-\cos h_{sr}]dx$
Polar tracking	$(ah_{\text{coll}}+b\sin h_{\text{coll}})/(d\cos L)$	$(h_{\text{coll}}/d)\{(1/\cos L)+[\cos h_{sr}/(\text{CR})]\}-\sin h_{\text{coll}}/[d(\text{CR})]$
Two-axis tracking	$(ah_{\text{coll}}+b\sin h_{\text{coll}})/(d\cos\delta_s\cos L)$	$(h_{\text{coll}}/d)(1/\cos\delta_s\cos L)+[\cos h_{sr}\cos L]+[\cos h_{sr}/(\text{CR})]-h_{\text{coll}}/[d(\text{CR})]$

[a] The collection hour angle value h_{coll} not used as the argument of trigonometric functions is expressed in radians; note that the total collection interval, $2h_{\text{coll}}$, is assumed to be centered about solar noon.

[b] $a = 0.409 + 0.5016\sin(h_{sr} - 60°)$.

[c] $c = 0.6609 - 0.4767\sin(h_{sr} - 60°)$.

[d] $d = \sin h_{sr} - h_{sr}\cos h_{sr}; \cos h_{sr}(i=90°) = -\tan\delta_s\tan(L-\beta)$.

[e] CR is the collector concentration ratio.

[f] Use elliptic integral tables to evaluate terms of the form of $\int_0^h\sqrt{\cos^2 x+\tan^2\delta_s}dx$ contained in r_T and r_d.

29

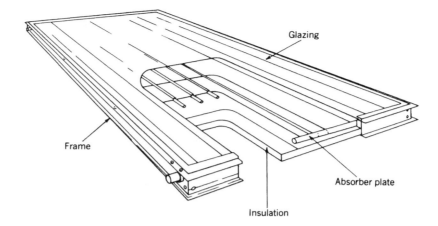

(a)

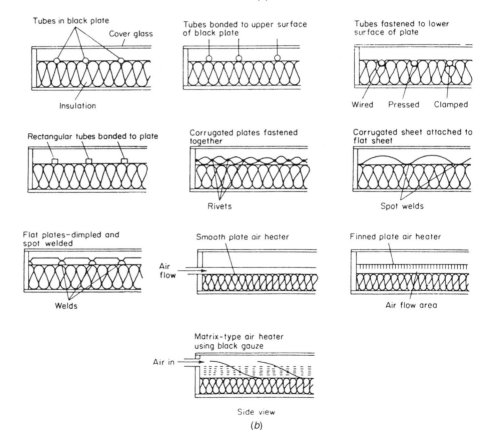

(b)

Figure 8 (a) Schematic diagram of solar collector with one cover. (b) Cross-sections of various liquid- and air-based flat-plate collectors in common use.

flat-plate collector. From the top down it contains a glazing system—normally one pane of glass, a dark-colored metal absorbing plate, insulation to the rear of the absorber, and, finally, a metal or plastic weatherproof housing. The glazing system is sealed to the housing to prohibit the ingress of water, moisture, and dust. The piping shown is thermally bonded to the absorber plate and contains the working fluid by which the heat produced is transferred to its end use. The pipes shown are manifolded together so that one inlet and one outlet connection, only, are present. Figure 8b shows a number of other collector designs in common use.

The energy produced by flat-plate collectors is the difference between the solar flux absorbed by the absorber plate and that lost from it by convection and radiation from the upper (or *front*) surface and that lost by conduction from the lower (or *back*) surface. The solar flux absorbed is the incident flux I_c Imultiplied by the glazing system transmittance τ and by the absorber plate absorptance α. The heat lost from the absorber in steady state is given by an overall thermal conductance U_c multiplied by the difference in temperature between the collector absorber temperature T_c and the surrounding, ambient temperature T_a. In equation form the net heat produced q_u is then

$$q_u = (\tau\alpha)I_c = U_c(T_c - T_0)(\text{W/m}^2) \tag{26}$$

The rate of heat production depends on two classes of parameters. The first—T_c, T_a, and I_c—having to do with the operational environment and the condition of the collector. The second—U_c and $\tau\alpha$—are characteristics of the collector independent of where or how it is used. The optical properties τ and α depend on the incidence angle, both dropping rapidly in value for $i > 50$–$55°$. The heat loss conductance can be calculated,[1] but formal tests, as subsequently described, are preferred for the determination of both $\tau\alpha$ and U_c.

Collector efficiency is defined as the ratio of heat produced q_u to incident flux I_c, that is,

$$\eta_c \equiv q_u/I_c \tag{27}$$

Using this definition with equation (26) gives the efficiency as

$$\eta_c = \tau\alpha - U_c\left(\frac{T_e - T_a}{I_c}\right) \tag{28}$$

The collector plate temperature is difficult to measure in practice, but the fluid inlet temperature $T_{f,i}$ is relatively easy to measure. Furthermore, $T_{f,i}$ is often known from characteristics of the process to which the collector is connected. It is common practice to express the efficiency in terms of $T_{f,i}$ instead of T_c for this reason. The efficiency is

$$\eta_c = F_k\left[\tau\alpha - U_c\left(\frac{T_{f,i} - T_a}{I_c}\right)\right] \tag{29}$$

in which the heat removal factor F_R is introduced to account for the use of $T_{j,i}$ for the efficiency basis. F_R depends on the absorber plate thermal characteristics and heat loss conductance.[2]

Equation (29) can be plotted with the group of operational characteristics $(T_{j,i} - T_a)/I_c$ as the independent variable as shown in Figure 9. The efficiency decreases linearly with the abscissa value. The intercept of the efficiency curve is the optical efficiency $\tau\alpha$ and the slope is $-F_R U_c$. Since the glazing transmittance and absorber absorptance decrease with solar incidence angle, the efficiency curve migrates toward the origin with increasing incidence angle, as shown in the figure. Data points from a collector test are also shown on the plot. The best-fit efficiency curve at normal incidence ($i = 0$) is determined numerically by a curve-fit method. The slope and intercept of the experimental curve, so determined, are the preferred values of the collector parameters as opposed to those calculated theoretically.

Selective Surfaces

One method of improving efficiency is to reduce radiative heat loss from the absorber surface. This is commonly done by using a low emittance (in the infrared

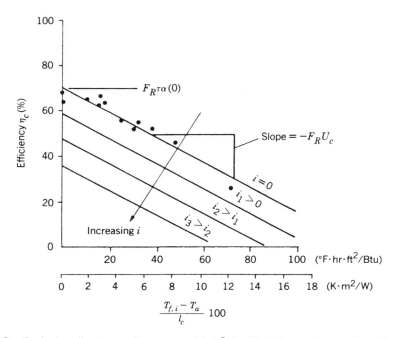

Figure 9 Typical collector performance with 0° incident beam flux angle. Also shown qualitatively is the effect of incidence angle i, which may be quantified by $\overline{\tau\alpha}(i)/\overline{\tau\alpha}(0) = 1.0 + b_0(1/\cos i - 1.0)$, where b_0 is the incidence angle modifier determined experimentally (ASHRAE 93-77) or from the Stokes and Fresnel equations.

Table 6   Selective Surface Properties

Material	Absorptance[a] α	Emittance ϵ	Comments
Black chrome	0.87–0.93	0.1	
Black zinc	0.9	0.1	
Copper oxide over aluminum	0.93	0.11	
Black copper over copper	0.85–0.90	0.08–0.12	Patinates with moisture
Black chrome over nickel	0.92–0.94	0.07–0.12	Stable at high temperatures
Black nickel over nickel	0.93	0.06	May be influenced by moisture
Black iron over steel	0.90	0.10	

region) surface having high absorptance for solar flux. Such surfaces are called (wavelength) *selective surface* and are used on very many flat-plate collectors to improve efficiency at elevated temperature. Table 6 lists emittance and absorptance values for a number of common selective surfaces. Black chrome is reliable and cost-effective.

2.2   Concentrating Collectors

Another method of improving the efficiency of solar collectors is to reduce the parasitic heat loss embodied in the second term of equation (29).[3] This can be done by reducing the size of the absorber relative to the aperture area. Relatively speaking, the area from which heat is lost is smaller than the heat collection area and efficiency increases. Collectors that focus sunlight onto a relatively small absorber can achieve excellent efficiency at temperatures above which flat-plate collectors produce no net heat output. In this section a number of concentrators are described.

Trough Collectors

Figure 10 shows cross-sections of five concentrators used for producing heat at temperatures up to 650°F at good efficiency. Figure 10*a* shows the parabolic "trough" collector representing the most common concentrator design available commercially. Sunlight is focused onto a circular pipe absorber located along the focal line. The trough rotates about the absorber centerline in order to maintain a sharp focus of incident beam radiation on the absorber. Selective surfaces and glass enclosures are used to minimize heat losses from the absorber tube.

Figures 10*c* and 10*d* show Fresnel-type concentrators in which the large reflector surface is subdivided into several smaller, more easily fabricated and shipped segments. The smaller reflector elements are easier to track and offer less wind resistance at windy sites; futhermore, the smaller reflectors are less costly. Figure 10*e* shows a Fresnel lens concentrator. No reflection is used with this approach; reflection is replaced by refraction to achieve the focusing effect. This device

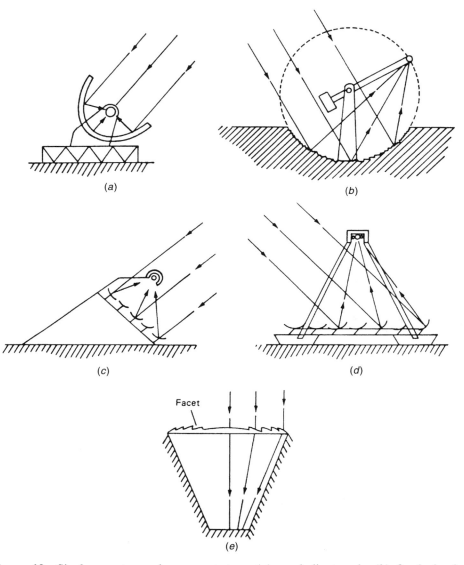

Figure 10 Single-curvature solar concentrators: (*a*) parabolic trough; (*b*) fixed circular trough with tracking absorber; (*c*) and (*d*) Fresnel mirror designs; and (*e*) Fresnel lens.

has the advantage that optical precision requirements can be relaxed somewhat relative to reflective methods.

Figure 10*b* shows schematically a concentrating method in which the mirror is fixed, thereby avoiding all problems associated with moving large mirrors to track the sun as in the case of concentrators described above. Only the absorber pipe is required to move to maintain a focus on the focal line.

The useful heat produced Q_u by any concentrator is given by

$$Q_u = A_d \eta_0 I_c - A_r U_c'(T_c - T_a) \tag{30}$$

in which the concentrator optical efficiency (analogous to $\tau \alpha$ for flat-plate collectors) is η_0, the aperture area is A_a, the receiver or absorber area is A_r, and the absorber heat loss conductance is U_c'. Collector efficiency can be found from equation (27) and is given by

$$\eta_c = \eta_0 - \frac{A_r}{A_a} U_c' \left(\frac{T_c - T_a}{I_c} \right) \tag{31a}$$

The aperture area-receiver area ratio $A_a/A_c > 1$ is called the *geometric concentration ratio* CR. It is the factor by which absorber heat losses are reduced relative to the aperture area:

$$\eta_c = \eta_0 - \frac{U_c'}{\mathrm{CR}} \left(\frac{T_c - T_a}{I_c} \right) \tag{31b}$$

As with flat-plate collectors, efficiency is most often based on collector fluid inlet temperature $T_{j,i}$ On this basis, efficiency is expressed as

$$\eta_c = F_R \left[\eta_0 - U_c \left(\frac{T_{j,i} - T_a}{I_c} \right) \right] \tag{32}$$

in which the heat loss conductance U_c on an aperture area basis is used ($U_c = U_c'/\mathrm{CR}$).

The optical efficiency of concentrators must account for a number of factors not present in flat-plate collectors including mirror reflectance, shading of aperture by receiver and its supports, spillage of flux beyond receiver tube ends at off-normal incidence conditions, and random surface, tracking, and construction errors that affect the precision of focus. In equation form the general optical efficiency is given by

$$\eta_0 = \rho_m \tau_c a_r \int_t \delta F(i) \tag{33}$$

where ρ_m is the mirror reflectance (0.8–0.9), τ_c is the receiver cover transmittance (0.85–0.92), α_r is the receiver surface absorptance (0.9–0.92), f_t is the fraction of aperture area r not shaded by receiver and its supports (0.95–0.97), δ is the intercept factor accounting for mirror surface and tracking errors (0.90–0.95), and $F(i)$ is the fraction of reflected solar flux intercepted by the receiver for perfect optics and perfect tracking. Values for these parameters are given in Refs. 2 and 4.

Compound Curvature Concentrators

Further increases in concentration and concomitant reductions in heat loss are achievable if dishtype concentrators are used. This family of concentrators is exemplified by the paraboloidal dish concentrator, which focuses solar flux at a point instead of along a line as with trough collectors. As a result the achievable

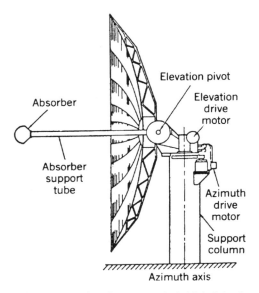

Figure 11 Segmented mirror approximation to paraboloidal dish. Average CR is 118, while maximum local CR is 350.

concentration ratios are approximately the square of what can be realized with single curvature, trough collectors. Figure 11 shows a paraboloidal dish concentrator assembly. These devices are of most interest for power production and some elevated industrial process heat applications.

For very large aperture areas it is impractical to construct paraboloidal dishes consisting of a single reflector. Instead, the mirror is segmented as shown in Figure 12. This collector system, called the *central receiver*, has been used in several solar thermal power plants in the 1- to 15-MW range. This power production method is discussed in the next section.

The efficiency of compound curvature dish collectors is given by equation (32), where the parameters involved are defined in the context of compound curvature optics.[4] The heat loss term at high temperatures achieved by dish concentrators is dominated by radiation; therefore, the second term of the efficiency equation is represented as

$$\eta_c = \eta_0 - \frac{\epsilon_i \sigma (T_\epsilon^4 - T_\alpha'^4)}{\text{CR}} \tag{34}$$

where ϵ_r the infrared emittance of the receiver, σ is the Stefan-Boltzmann constant, and T_n' is the equivalent ambient temperature for radiation, depending on ambient humidity and cloud cover. For clear, dry conditions T_n^θ is about 15°F to 20°F below the ambient dry-bulb temperature. As humidity decreases, T_θ approaches the dry-bulb temperature.

The optical efficiency for the central receiver is expressed in somewhat different terms than those used in equation (33). It is referenced to solar flux on a

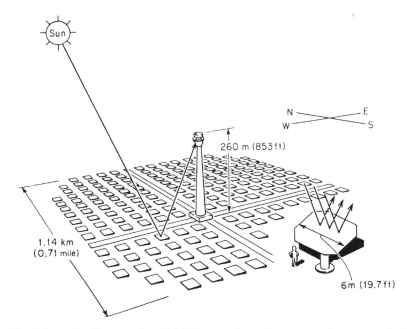

Figure 12   Schematic diagram of a 50-MWe central receiver power plant. A single heliostat is shown in the inset to indicate its human scale. (From Ref. 5.)

horizontal surface and therefore includes the geometric tilt factor. For the central receiver, the optical efficiency is given by

$$\eta_0 = \phi \mathscr{G} \rho_{\tau R} \alpha_r \, f_t \delta \tag{35}$$

in which the last four parameters are defined as in equation (33). The ratio of redirected flux to horizontal flux is $\mathscr{G}$ and is given approximately by

$$\mathscr{G} = 0.78 + 1.5(1 - \alpha/90)^2 \tag{36}$$

from Ref. 4. The ratio of mirror area to ground area ϕ depends on the size and economic factors applicable to a specific installation. Values for ϕ have been in the range 0.4 to 0.5 for installations made through 1985.

2.3   Collector Testing

To determine the optical efficiency and heat-loss characteristics of flat-plate and concentrating collectors (other than the central receiver, which is difficult to test because of its size), testing under controlled conditions is preferred to theoretical calculations. Such test data are required if comparisons among collectors are to be made objectively. As of the mid-1980s very few consensus standards had been adopted by the U.S. solar industry. The ASHRAE Standard Number 93-77 applies to flat-plate collectors that contain either a liquid or a gaseous working fluid.[6] Collectors in which a phase change occurs are not included. In addition, the standards do not apply well to concentrators, since additional procedures are needed to find the optical efficiency and aging effects. Testing of concentrators uses sections of the above standard where applicable plus additional procedures as needed; however, no industry standard exists. (The ASTM has promulgated standard E905 as the first proposed standard for concentrator tests.) ASHRAE Standard Number 96-80 applies to very-low-temperature collectors manufactured without any glazing system.

Figure 13 shows the test loop used for liquid-cooled flat-plate collectors. Tests are conducted with solar flux at near-normal incidence to find the normal incidence optical efficiency $(\tau\alpha)_n$ along with the heat loss conductance U_e. Off-normal optical efficiency is determined in a separate test by orienting the collector such that several substantially offnormal values of $\tau\alpha$ or η_0 can be measured. The fluid used in the test is preferably that to be used in the installed application, although this is not always possible. If operational and test fluids differ, an analytical correction in the heat removal factor FR is to be made.[2] An additional test is made after a period of time (nominal one month) to determine

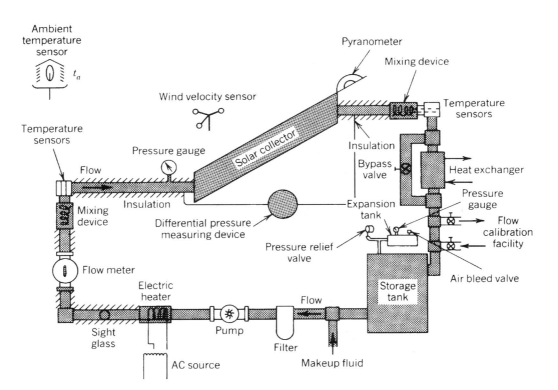

Figure 13   Closed-loop testing configuration for the solar collector when the transfer fluid is a liquid.

the effect of aging, if any, on the collector parameters listed above. A similar test loop and procedure apply to air-cooled collectors.[6]

The development of full system tests has only begun. Of course, it is the entire solar system (see next section) not just the collector that ultimately must be rated in order to compare solar and other energy-conversion systems. Testing of full-size solar systems is very difficult owing to their large size and cost. Hence, it is unlikely that full system tests will ever be practical except for the smallest systems such as residential water heating systems. For this one group of systems a standard test procedure (ASHRAE 95-81) exists. Larger-system performance is often predicted, based on component tests, rather than measured.

3 SOLAR THERMAL APPLICATIONS

One of the unique features of solar heat is that it can be produced over a very broad range of temperatures—the specific temperature being selected to match the thermal task to be performed. In this section the most common thermal applications will be described in summary form. These include low-temperature uses such as water and space heating ($30°–100°C$), intermediate temperature industrial processes ($100°–300°C$), and high-temperature thermal power applications ($500°–850°C$ and above). Methods for predicting performance, where available, will also be summarized. Nonthermal solar applications are described in the next section.

3.1 Solar Water Heating

The most often used solar thermal application is for the heating of water for either domestic or industrial purposes.[7] Relatively simple systems are used, and the load exists relatively uniformly through a year resulting in a good system load factor. Figure 14a shows a singletank water heater schematically. The key components are the collector ($0.5–1.0$ ft^2/gal day load), the storage tank ($1.0–2.0$ gal/ft^2 of collector), a circulating pump, and controller. The check valve is essential to prevent backflow of collector fluid, which can occur at night when the pump is off if the collectors are located some distance above the storage tank. The controller actuates the pump whenever the collector is $15°$F to $30°$F warmer than storage. Operation continues until the collector is only $1.5°$F to $5°$F warmer than the tank, at which point it is no longer worthwhile to operate the pump to collect the relatively small amounts of solar heat available.

The water-heating system shown in Figure 14a uses an electrical coil located near the top of the tank to ensure a hot water supply during periods of solar outage. This approach is only useful in small residential systems and where nonsolar energy resources other than electricity are not available. Most commercial systems are arranged as shown in Figure 14b, where a separate preheat tank, heated only by solar heat, is connected upstream of the nonsolar, auxiliary water heater

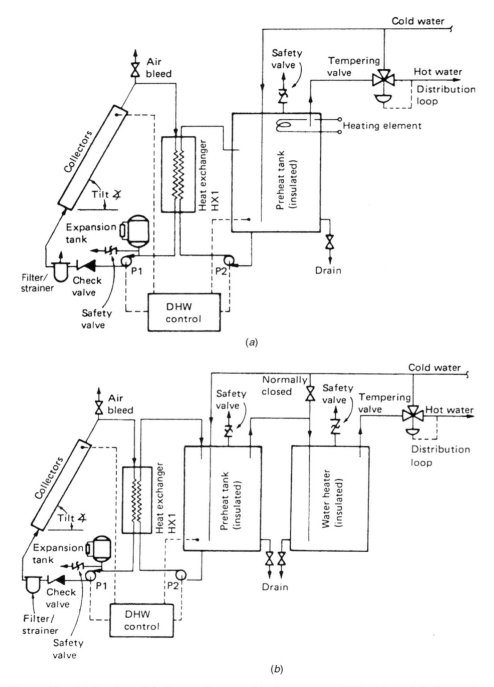

Figure 14   (*a*) Single-tank indirect solar water-heating system. (*b*) Double-tank indirect solar water-heating system. Instrumentation and miscellaneous fittings are not shown.

tank or plant steam-to-water heat exchanger. This approach is more versatile in that any source of backup energy whatever can be used when solar heat is not available. Additional parasitic heat loss is encountered, since total tank surface area is larger than for the single tank design.

The water-heating systems shown in Figure 14 are of the indirect type, that is, a separate fluid is heated in the collector and heat thus collected is transferred to the end use via a heat exchanger. This approach is needed in locations where freezing occurs in winter and antifreeze solutions are required. The heat exchanger can be eliminated, thereby reducing cost and eliminating the unavoidable fluid temperature decrement between collector and storage fluid streams, if freezing will never occur at the application site. The exchanger can also be eliminated if the *drain-back approach* is used. In this system design, the collectors are filled with water only when the circulating pump is on—that is, only when the collectors are warm. If the pump is not operating, the collectors and associated piping all drain back into the storage tank. This approach has the further advantage that heated water otherwise left to cool overnight in the collectors is returned to storage for useful purposes.

The earliest water heaters did not use circulating pumps, but used the density difference between cold collector inlet water and warmer collector outlet water to produce the flow. This approach is called a *thermosiphon* and is shown in Figure 15. These systems are among the most efficient, since no parasitic use of electric pump power is required. The principal difficulty is the requirement that the large storage tank be located above the collector array, often resulting in structural and architectural difficulties. Few industrial solar water-heating systems have used this approach, owing to difficulties in balancing buoyancy-induced flows in large piping networks.

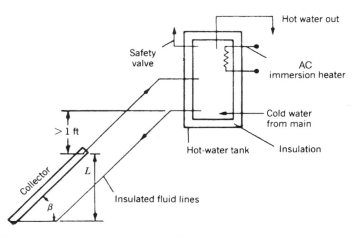

Figure 15 Passive thermosiphon single-tank direct system for solar water heating. Collector is positioned below the tank to avoid reverse circulation.

3.2   Mechanical Solar Space Heating Systems

Solar space heating is accomplished using systems similar to those for solar water heating. The collectors, storage tank, pumps, heat exchangers, and other components are larger in proportion to the larger space heat loads to be met by these systems in building applications. Figure 16 shows the arrangement of components in one common space heating system. All components except the solar collector and controller have been in use for many years in building systems and are not of special design for the solar application.

The control system is somewhat more complex than that used in nonsolar building heating systems, since two heat sources—solar and nonsolar auxiliary—are to be used under different conditions. Controls using simple microprocessors are available for precise and reliable control of solar space-heating systems.

Air-based systems are also widely used for space heating. They are similar to the liquid system shown in Figure 16 except that no heat exchanger is used and

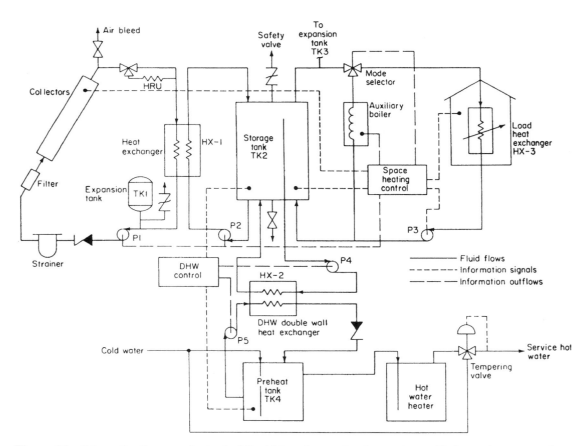

Figure 16   Schematic diagram of a typical liquid-based space-heating system with domestic water preheat.

rock piles, not tanks of fluid, are the storage media. Rock storage is essential to efficient air-system operation since gravel (usually 1–2 in. in diameter) has a large surface-to-volume ratio necessary to offset the poor heat transfer characteristics of the air working fluid. Slightly different control systems are used for air-based solar heaters.

3.3 Passive Solar Space Heating Systems

An effective way of heating residences and small commercial buildings with solar energy and without significant nonsolar operating energy is the *passive* heating approach.[1] Solar flux is admitted into the space to be heated by large, sun-facing apertures. In order that overheating not occur during sunny periods, large amounts of thermal storage are used, often also serving a structural purpose. A number of classes of passive heating systems have been identified and are described in this section.

Figure 17 shows the simplest type of passive system known as *direct gain*. Solar flux enters a large aperture and is converted to heat by absorption on dark colored floors or walls. Heat produced at these wall surfaces is partly conducted into the wall or floor serving as stored heat for later periods without sun. The remaining heat produced at wall or floor surfaces is convected away from the surface thereby heating the space bounded by the surface. Direct-gain systems also admit significant daylight during the day; properly used, this can reduce artificial lighting energy use. In cold climates significant heat loss can occur through the solar aperture during long, cold winter nights. Hence, a necessary component of efficient direct-gain systems is some type of insulation system put in place at night over the passive aperture. This is indicated by the dashed lines in Figure 17*b*.

The second type of passive system commonly used is variously called the thermal storage wall (TSW) or collector storage wall. This system, shown in Figure 18, uses a storage mass interposed between the aperture and space to be heated. The reason for this positioning is to better illuminate storage for a significant part of the heating season and also to obviate the need for a separate insulation system; selective surfaces applied to the outer storage wall surface are able to control heat loss well in cold climates, while having little effect on solar absorption. As shown in the figure, a thermocirculation loop is used to transport heat from the warm, outer surface of the storage wall to the space interior to the wall. This air flow convects heat into the space during the day, while conduction through the wall heats the space after sunset. Typical storage media include masonry, water, and selected eutectic mixtures of organic and inorganic materials. The storage wall eliminates glare problems associated with direct-gain systems, also.

The third type of passive system in use is the attached greenhouse or sunspace, as shown in Figure 19. This system combines certain features of both direct-gain

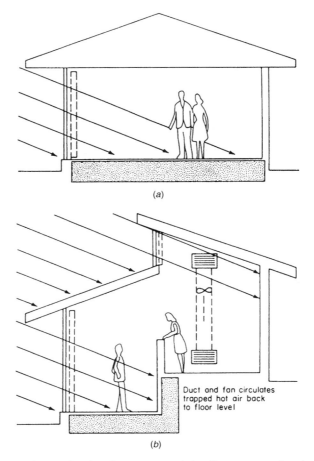

Duct and fan circulates
trapped hot air back
to floor level

(b)

Figure 17 Direct-gain passive heating systems: (a) adjacent space heating; (b) clerestory for north zone heating.

and storage wall systems. Night insulation may or may not be used, depending on the temperature control required during nighttime.

The key parameters determining the effectiveness of passive systems are the optical efficiency of the glazing system, the amount of directly illuminated storage and its thermal characteristics, the available solar flux in winter, and the thermal characteristics of the building of which the passive system is a part. In a later section, these parameters will be quantified and will be used to predict the energy saved by the system for a given building in a given location.

3.4 Solar Ponds

A *solar pond* is a body of water no deeper than a few meters configured in such a way that usual convection currents induced by solar absorption are suppressed.[6]

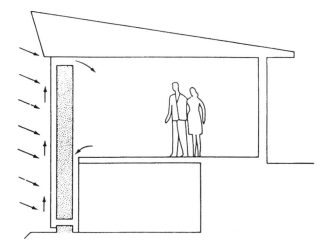

Figure 18 Indirect-gain passive system—TSW system.

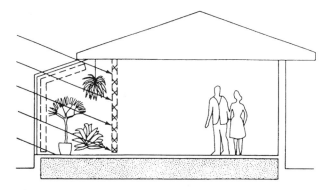

Figure 19 Greenhouse or attached sun-space passive heating system using a combination of direct gain into the greenhouse and indirect gain through the thermal storage wall, shown by cross-hatching, between the greenhouse and the living space.

The oldest method for convection suppression is the use of high concentrations of soluble salts in layers near the bottom of the pond with progressively smaller concentrations near the surface. The surface layer itself is usually fresh water. Incident solar flux is absorbed by three mechanisms. Within a few millimeters of the surface the infrared component (about one-third of the total solar flux energy content) is completely absorbed. Another third is absorbed as the visible and ultraviolet components traverse a pond of nominal 2 m depth. The remaining one-third is absorbed at the bottom of the pond. It is this component that would induce convection currents in a freshwater pond, thereby causing warm water to rise to the top where convection and evaporation would cause substantial heat loss. With proper concentration gradient, convection can be completely suppressed and significant heat collection at the bottom layer is possible. Salt gradient ponds are hydrodynamically stable if the following criterion is satisfied:

$$\frac{d\rho}{dz} = \frac{\partial \rho}{\partial s}\frac{ds}{dz} + \frac{\partial \rho}{\partial T}\frac{dT}{dz} > 0 \tag{37}$$

where s is the salt concentration, ρ is the density, T is the temperature, and z is the vertical coordinate measured positive downward from the pond surface. The inequality requires that the density must decrease upward.

Useful heat produced is stored in and removed from the lowest layer as shown in Figure 20. This can be done by removing the bottom layer of fluid, passing it through a heat exchanger, and returning the cooled fluid to another point in the bottom layer. Alternatively, a network of heat-removal pipes can be placed on the bottom of the bond and the working fluid passed through for heat collection. Depending on the design, solar ponds also may contain substantial heat storage capability if the lower convective zone is relatively thick. This approach is used when uniform heat supply is necessary over a 24 hour period but solar flux is available for only a fraction of the period. Other convection-suppression techniques and heat-removal methods have been proposed but not used in more than one installation at most.

The requirements for an effective solar pond installation include the following. Large amounts of nearly free water and salt must be available. The subsoil must be stable in order to minimize changes in pond shape that could fracture the waterproof liner. Adequate solar flux is required year around; therefore, pond usage is confined to latitudes within $40°$ of the equator. Fresh-water aquifers used for potable water should not be nearby in the event of a major leak of saline water into the groundwater. Other factors include low winds to avoid surface waves and windblown dust collection within the pond (at the neutral buoyancy point),

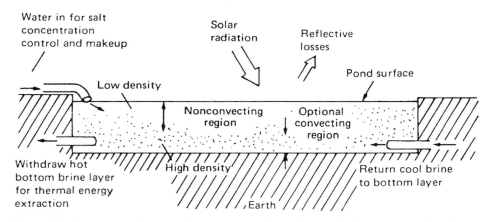

Figure 20 Schematic diagram of a nonconvecting solar pond showing conduits for heat withdrawal, surface washing, and an optional convecting zone near the bottom.

low soil conductivity (i.e., low water content) to minimize conduction heat loss, and durable liner materials capable of remaining leakproof for many years.

The principal user of solar ponds has been the country of Israel. Ponds tens of acres in size have been built and operated successfully. Heat collected has been used for power production with an organic Rankine cycle, for space heating, and for industrial uses. A thorough review of solar pond technology is contained in Ref. 6. A method for predicting the performance of a solar pond is presented in the next section.

3.5   Solar Thermal Power Production

Solar energy has very high thermodynamic availability owing to the high effective temperature of the source. Therefore, production of shaft power and electric power therefrom is thermodynamically possible. Two fundamentally different types of systems can be used for power production: (1) a large array of concentrating collectors of several tens of meters in area connected by a fluid or electrical network and (2) a single, central receiver using mirrors distributed over a large area but producing heat and power only at one location. The determination of which approach is preferred depends on required plant capacity. For systems smaller than 10 MW the distributed approach appears more economical with existing steam turbines. For systems greater than 10 MW, the central receiver appears more economical.[8] However, if highly efficient Brayton or Stirling engines were available in the 10–20 kW range, the distributed approach would have lowest cost for any plant size. Such systems will be available by the year 2000.

The first U.S. central receiver began operating in the fall of 1982. Located in the Mojave Desert, this 10-MW plant (called *Solar One*) is connected to the southern California electrical grid. The collection system consists of 1818 heliostats totaling 782,000 ft^2 in area. Each 430 ft^2 mirror is computer controlled to focus reflected solar flux onto the receiver located 300 ft above the desert floor. The receiver is a 23-ft diameter cylinder whose outer surface is the solar absorber. The absorbing surface is coated with a special black paint selected for its reliability at the nominal 600°C operating temperature. Thermal storage consisting of a mixture of an industrial heat transfer oil for heat transport and of rock and sand has a nominal operating temperature of 300°C. Storage is used to extend the plant operating time beyond sunset (albeit at lower turbine efficiency) and to maintain the turbine, condenser, and piping at operating temperatures overnight as well as to provide startup steam the following morning. The plant was modernized in 1996 and operated for several more years.

Solar-produced power is not generally cost-effective currently. The principal purpose of the Solar One experiment and other projects in Europe and Japan is to acquire operating experience with the solar plant itself as well as with the interaction of solar and nonsolar power plants connected in a large utility grid.

Extensive data collection and analysis will answer questions regarding long-term net efficiency of solar plants, capacity displacement capability, and reliability of the new components of the system—mirror field, receiver, and computer controls.

3.6   Other Thermal Applications

The previous sections have discussed the principal thermal applications of solar energy that have been reduced to practice in at least five different installations and that show significant promise for economic displacement of fossil or fissile energies. In this section two other solar-conversion technologies are summarized.

Solar-powered cooling has been demonstrated in many installations in the United States, Europe, and Japan. Chemical absorption, organic Rankine cycle, and desiccant dehumidifaction processes have all been shown to be functional. Most systems have used flat-plate collectors, but higher coefficients of performance are achievable with mildly concentrating collectors. Reference 7 describes solar-cooling technologies. To date, economic viability has not been generally demonstrated, but further research resulting in reduced cost and improved efficiency is expected to continue.

Thermal energy stored in the surface layers of the tropical oceans has been used to produce electrical power on a small scale. A heat engine is operated between the warmest layer at the surface and colder layers several thousand feet beneath. The available temperature difference is of the order of $20°$ C, therefore, the cycle efficiency is very low—only a few percent. However, this type of power plant does not require collectors or storage. Only a turbine capable of operating efficiently at low temperature is needed. Some cycle designs also require very large heat exchangers, but new cycle concepts without heat exchangers and their unavoidable thermodynamic penalties show promise.

3.7   Performance Prediction for Solar Thermal Processes

In a rational economy the single imperative for use of solar heat for any of the myriad applications outlined heretofore must be cost competitiveness with other energy sources—fossil and fissile. The amount of useful solar energy produced by a solar-conversion system must therefore be known, along with the cost of the system. In this section the methods usable for predicting the performance of widely deployed solar systems are summarized. Special systems such as the central receiver, the ocean thermal power plant, and solar cooling are not included. The methods described here require a minimum of computational effort, yet embody all important parameters determining performance.

Solar systems are connected to end uses characterized by an energy requirement or "load" L and by operating temperature that must be achievable by the solar-heat-producing system. The amount of solar-produced heat delivered to the end use is the useful energy Q_u. This is the net heat delivery accounting for parasitic losses in the solar subsystem. The ratio of useful heat delivered to the requirement L is called the *solar fraction* denoted by f_s. In equation form the solar fraction is

$$f_s = \frac{Q_u}{L} \tag{38}$$

Empirical equations have been developed relating the solar fraction to other dimensionless groups characterizing a given solar process. These are summarized shortly.

A fundamental concept used in many predictive methods is the solar *utilizability* defined as that portion of solar flux absorbed by a collector that is capable of providing heat to the specified end use. The key characteristic of the end use is its temperature. The collector must produce at least enough heat to offset losses when the collector is at the minimum temperature $T_{\min}$ usable by the given process. Figure 21 illustrates this idea schematically. The curve represents the flux absorbed over a day by a hypothetical collector. The horizontal line intersecting this curve represents the threshold flux that must be exceeded for a net energy collection to take place. In the context of the efficiency equation (32), this critical flux I_{cr} is that which results in a collector efficiency of exactly zero when the collector is at the minimum usable process temperature $T_{\min}$. Any greater flux will result in net heat production. From equation (32) the critical intensity is

$$I_{cr} = \frac{U_c(T_{\min} - T_\alpha)}{T\alpha} \tag{39}$$

The solar utilizability is the ratio of the useful daily flux (area above I_{cr} line in Figure 21) to the total absorbed flux (area $A_1 + A_2$) beneath the curve. The utilizability denoted by ϕ is

$$\phi = \frac{A_1}{A_1 + A_2} \tag{40}$$

This quantity is a solar radiation statistic depending on I_{cr}, characteristics of the incident solar flux and characteristics of the collection system. It is a very useful parameter in predicting the performance of solar thermal systems.

Table 7 summarizes empirical equations used for predicting the performance of the most common solar-thermal systems. These expressions are given in terms of the solar fraction defined above and dimensionless parameters containing all important system characteristics. The symbols used in this table are defined in Table 8. In the brief space available in this chapter, all details of these prediction methodologies cannot be included. The reader is referred to Refs. 1, 3, 8, and 9 for details.

4 PHOTOVOLTAIC SOLAR ENERGY APPLICATIONS

In this section the principal nonthermal solar conversion technology is described. Photovoltaic cells are capable of converting solar flux directly into electric power. This process, first demonstrated in the 1950s, holds considerable promise for

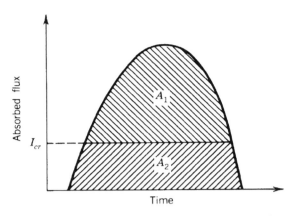

Figure 21 Daily absorbed solar flux ($A_1 + A_2$) and useful solar flux (A_1) at intensities above I_{cr}.

Table 7 Empirical Solar Fraction Equations[a]

System Type	f_s Expression	Time Scale
Water heating and liquid-based space heating	$f_s = 1.029 P_z s - 0.065 P_L - 0.245 P_s^2 + 0.0018 P_L^2 + 0.00215 P_s^3$	Monthly
Space heating—air-based systems	$f_s = 1.040 P_s - 0.065 P_L - 0.159 P_s^2 + 0.00187 P_L^2 + 0.0095 P_s^3$	Monthly
Passive direct gain	$f_s = PX + (1 - P)(3.082 - 3.142\overline{\phi})(1 - e^{-0.329x})$	Monthly
Passive storage wall	$f_s = Pf_\infty + 0.88(1 - P)(1 - e^{-1.26f_\infty})$	Monthly
Concentrating collector systems	$f_s = F_R \overline{\eta}_0 \overline{I}_c A_c N \overline{\phi}' / L$	Monthly
Solar ponds (pond radius R to provide annual pond temperature T_p)	$R = \dfrac{2.2\overline{\Delta T} + [4.84(\Delta T)^2 + \overline{L}(0.3181\overline{I}_p - 0.1592\Delta T)]^{1/2}}{\overline{I}_p - 0.5\Delta T}$	Annual

[a] See Table 8 for symbol definitions.

significant use in the future. Major cost reductions have been accomplished. In this section the important features of solar cells are described.

Photovoltaic conversion of sunlight to electricity occurs in a thin layer of semiconductor material exposed to solar flux. Photons free electric charges, which flow through an external circuit to produce useful work. The semiconductor materials used for solar cells are tailored to be able to convert the majority of terrestrial solar flux; however, low-energy photons in the infrared region are usually not usable. Figure 22 shows the maximum theoretical conversion efficiency

Table 8 Definition of Symbols in Table 7

Parameters			Definition	Units[a]
P_L			$P_s = \dfrac{F_{hx}F_R U_c(T_r - \overline{T}_a)\Delta t}{L}$	None
	F_{hx}		$F_{hx} = \left\{\left[1 + \dfrac{F_R U_c A_c}{(\dot{m}C_p)_c}\right]\left[\dfrac{(\dot{m}C_p)_c}{(\dot{m}C_p)_{\min}\epsilon} - 1\right]\right\}^{-1},$ collector heat exchanger penalty factor	None
		$F_R U_c$	Collector heat-loss conductance	But/hr·ft^2 · °F
		A_c	Collector area	ft^2
		$(\dot{m}C_p)_c$	Collector fluid capacitance rate	Btu/hr·°F
		$(\dot{m}C_p)_{\min}$	Minimum capacitance rate in collector heat exchanger	Btu/hr·°F
		ϵ	Collector heat-exchanger effectiveness	None
		T_r	Reference temperature, 212°F	°F
		$\overline{T}_a$	Monthly averaged ambient temperature	°F
		Δt	Number of hours per month	hr/month
		L	Monthly load	Btu/month
P_s			$P_s = \dfrac{F_{hx}F_R \overline{\tau\alpha}\overline{I}_c N}{L}$	None
		$F_R\overline{T\alpha}$	Monthly averaged collector optical efficiency	None
		$\overline{I}_c$	Monthly averaged, daily incident solar flux	Btu/day·ft^2
		N	Number of days per month	day/month
(P'_L—to be used for water heating only)			$P'_L = P_L \dfrac{(1.18T_{wo}+3.86T_{wi}-2.32\overline{T}_a-66.2)}{212-\overline{T}_a}$	None
	T_{wo}		Water output temperature	°F
	T_{wi}		Water supply temperature	°F
	P_L		(See above)	None
P			$P = (1 - e^{-0.294Y})^{0.652}$	None
	Y		Storage-vent ratio, $Y = \dfrac{C\Delta T}{\phi \overline{I}_c\overline{\tau\alpha}A_c}$	None
	C		Passive storage capacity	Btu/°F
	ΔT		Allowable diurnal temperature saving in heated space	°F
$\overline{\phi}$			Monthly averaged utilizability (see below)	None
X			Solar-load ratio, $X = \dfrac{\overline{I}_c\overline{\tau\alpha}A_c N}{L}$	None
	L		Monthly space heat load	Btu/month
f_∞			Solar fraction with hypothetically infinite storage, $f_\infty = \dfrac{\overline{Q}_i + L_w}{L}$	None
	$\overline{Q}_i$		Net monthly heat flow through storage wall from outer surface to heated space	Btu/month
	L_w		Heat *loss* through storage wall	Btu/month

Table 8   (*continued*)

Parameters		Definition	Units[a]
$F_R \overline{\eta}_0$		Monthly averaged concentrator optical efficiency	None
$\overline{\phi}'$		Monthly average utilizability for concentrators	None
R		Pond radius to provide diurnal average pond temperature $\overline{T}_p$	m
ΔT		$\overline{\Delta T} = \overline{T}_p - \overline{\overline{T}}_a$	°C
	$\overline{T}_p$	Annually averaged pond temperature	°C
	$\overline{\overline{T}}_a$	Annually averaged ambient temperature	°C
$\overline{L}$		Annual averaged load at $\overline{T}_p$	W
$\overline{I}_p$		Annual averaged insolation absorbd at pond bottom	W/m²
$\overline{\phi}$		Monthly flat-plate utilizability (equator facing collectors), $\overline{\phi} = \exp\{[A + B(\overline{R}_N/\overline{R})](\overline{X}_c + C\overline{X}_c^2)\}$	None
	A	$A = 7.476 - 20.0\overline{K}_T + 11.188\overline{K}_T^2$	None
	B	$B = -8.562 + 18.679\overline{K}_T - 9.948\overline{K}_T^2$	None
	C	$C = -0.722 + 2.426\overline{K}_T + 0.439\overline{K}_T^2$	None
	$\overline{R}$	Tilt factor, see Eq. (21)	None
	$\overline{R}_N$	Monthly averaged tilt factor for hour centered about noon (see Ref. 10)	None
	$\overline{X}_c$	Critical intensity ratio, $\overline{X}_c = \frac{I_{cr}}{r_{T,N}\overline{R}_N \overline{H}_h}$	None
	$r_{T,N}$	Fraction of daily total radiation contained in hour about noon, $r_{T,N} = r_{d,n}[1.07 + 0.025\sin(h_{sr} - 60)]$	day/hr
		$r_{d,n} = \frac{\pi}{24}\frac{1-\cos h_{sr}}{h_{sr} - h_{sr}\cos h_{sr}}$	day/hr
	I_{cr}	Critical intensity [see Eq. (39)]	Btu/hr·ft²
$\overline{\phi}'$		Monthly concentrator utilizability, $\overline{\phi}' = 1.0 - (0.049 + 1.49\overline{K}_T)\overline{X}$ $+0.341\overline{K}_T\overline{X}^2 0.0 < \overline{K}_T < 0.75, 0 < \overline{X} < 1.2)$ $\overline{\phi}' = 1.0 - \overline{X}(\overline{K}_T)0.75, 0 < \overline{X} < 1.0)$	None
	$\overline{X}$	Concentrator critical intensity ratio, $\overline{X} = \frac{U_c(T_{f,i} - \overline{T}_a)\Delta t_c}{\overline{\eta}_0 \overline{I}_c}$	None
	$T_{j,i}$	Collector fluid inlet temperature—assumed constant	°F
	Δt_c	Monthly averaged solar system operating time	hr/day

[a]USCS unit shown except for solar ponds; SI units may also be used for all parameters shown in USCS units.

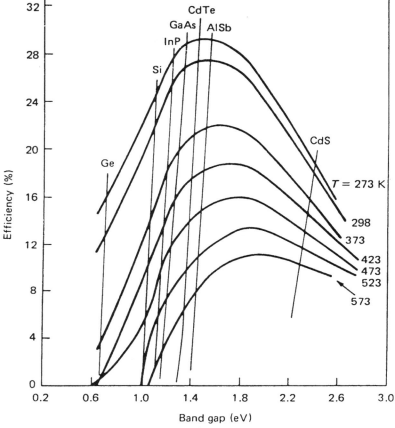

Figure 22 Maximum theoretical efficiency of photovoltaic converters as a function of band-gap energy for several materials.

of seven common materials used in the application. Each material has its own threshold band-gap energy, which is a weak function of temperature. The energy contained in a photon is $E = hv$. If E is greater than the band-gap energy shown in this figure, conversion can occur.

Figure 22 also shows the very strong effect of temperature on efficiency. For practical systems it is essential that the cell be maintained as near to ambient temperature as possible.

Solar cells produce current proportional to the solar flux intensity with wavelengths below the band-gap threshold. Figure 23 shows the equivalent circuit of a solar cell. Both internal shunt and series resistances must be included. These result in unavoidable parasitic loss of part of the power produced by the equivalent circuit current source of strength I_S. Solving the equivalent circuit for the

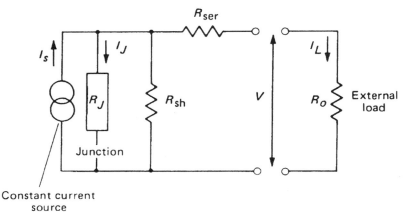

Figure 23 Equivalent circuit of an illuminated *p-n* photocell with internal series and shunt resistances and nonlinear junction impedance R_J.

power P produced and using an expression from Ref. 1 for the junction leakage I_j results in

$$P = [I - I_0(e^{p_0/kT} - 1)]V \qquad (41)$$

in which e_0 is the electron charge, k is the Boltzmann constant, and T is the temperature. The current source I_s is given by

$$I_s = \eta_0(1 - \rho_0)\alpha e_0 n_p \qquad (42)$$

in which η_0 is the collector carrier efficiency, ρ_0 is the cell surface reflectance, α is the absorptance of photons, and n_p is the flux density of sufficiently energetic photons.

In addition to the solar cell, complete photovoltaic systems also must contain electrical storage and a control system. The cost of storage presents another substantial cost problem in the widespread application of photovoltaic power production. The costs of the entire conversion system must be reduced by an order of magnitude in order to be competitive with other power sources. Vigorous research in the United States, Europe, and Japan has made significant gains in the past decade.

Figure 24 shows the effects of illumination intensity and cell temperature. Temperature affects the performance in a way that the voltage and thus the power output decrease with increasing temperature.

Efficiency is given by the ratio of useful output to insolation input:

$$n = E/(A_0 I_0) \qquad (43)$$

E is the area of the maximum rectangle inscribable within the IV curve shown in the figure, part *a*. A_c is the collector area and I_c is the incoming collector plane insolation. The characteristics noted in the IV curves below show that

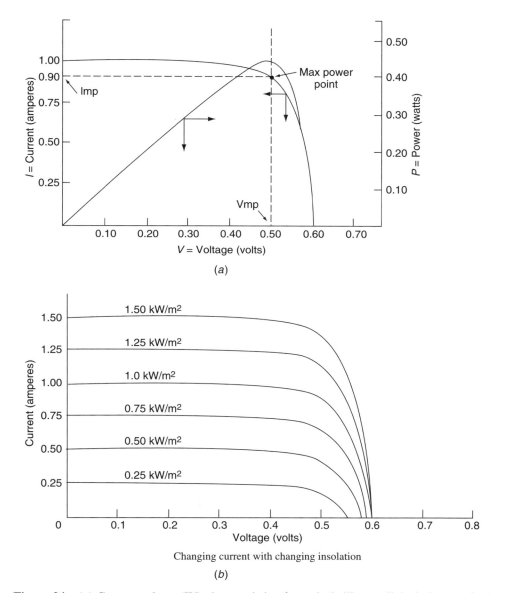

Figure 24   (*a*) Current-voltage (IV) characteristic of a typical silicon cell depicting nominal power output as the area of the maximal rectangle that can be inscribed within the IV curve; (*b*) typical IV curves showing the effects of illumination level and (*c*) the effects of cell temperature.

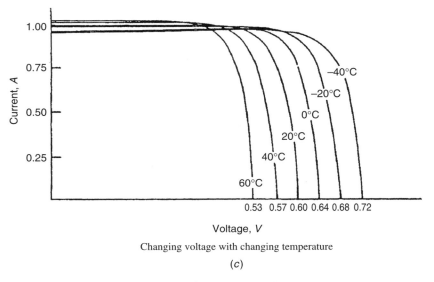

Changing voltage with changing temperature

(c)

Figure 24 (*Continued*)

efficiency is essentially independent of insolation and varies inversely with cell temperature. A useful efficiency equation is

$$n = n_R(1 - \beta[T_c - T_R])$$ (44)

$n_R =$ is the reference efficiency (mfr's data)
$\beta =$ is the temperature coefficient (0.004/°C)
where $T_C =$ is the PV cell temperature (°C)
$T_R =$ is the cell reference temperature at which the efficiency is n_R (°C)

This expression can be used with weather data to find the electrical output E from this expression, which follows from the definition of efficiency

$$E = n A_c I_c = \{n_R(1 - \beta[T_C - T_R])\} A_c I_c$$

Typical efficiencies for crystalline solar cells are 15 to 20 percent at rated conditions.

REFERENCES

1. Y. Goswami, F. Kreith, and J. F. Kreider, *Principles of Solar Engineering*, 2nd ed., Taylor & Francis, New York, 2000.
2. M. Collares-Pereira and A. Rabl, "The Average Distribution of Solar Energy," *Solar Energy* **22**, 155–164 (1979).
3. J. F. Kreider, *Medium and High Temperature Solar Processes*, Academic Press, New York, 1979.

4. ASHRAE Standard 93-77, "Methods of Testing to Determine the Thermal Performance of Solar Collectors," ASHRAE, Atlanta, GA, 1977.

5. Electric Power Research Institute (EPRT).

6. H. Tabor, "Solar Ponds— Review Article," *Solar Energy* **27**, 181–194 (1981).

7. J. F. Kreider and F. Kreith, *Solar Heating and Cooling*, Hemisphere/McGraw-Hill, New York, 1982.

8. M. Collares-Pereira and A. Rabl, "Simple Procedure for Predicting Long Term Average Performance of Nonconcentrating and of Concentrating Solar Collectors," *Solar Energy* **23**, 235–253 (1979).

9. J. F. Kreider, *The Solar Heating Design Process*, McGraw-Hill, New York, 1982.

CHAPTER 3

FUEL CELLS

Matthew M. Mench
Department of Mechanical and Nuclear Engineering
The Pennsylvania State University
University Park, Pennsylvania

1	**INTRODUCTION**	**59**
2	**BASIC OPERATING PRINCIPLES, EFFICIENCY, AND PERFORMANCE**	**64**
	2.1 Description of a Fuel-Cell Stack	67
	2.2 Performance and Efficiency Characterization	68
	2.3 Polarization Curve	72
	2.4 Heat Management	75
	2.5 Degradation	76
	2.6 Hydrogen PEFC	78
	2.7 H_2 PEFC Performance	78
	2.8 Technical Issues in H_2 PEFC	78
	2.9 Direct Methanol Fuel Cell	83
	2.10 Technical Issues of the DMFC	83
3	**SOLID OXIDE FUEL CELL**	**85**
	3.1 Technical Issues of SOFC	86
	3.2 Performance and Materials	86
4	**OTHER FUEL CELLS**	**91**
	4.1 Alkaline Fuel Cells	91
	4.2 Molten Carbonate Fuel Cells	91
	4.3 Phosphoric Acid Fuel Cells	94
	4.4 Other Alternatives	95
	NOMENCLATURE	**95**

1 INTRODUCTION

In 1839, Sir William Grove conducted the first known demonstration of the fuel cell. It operated with separate platinum electrodes in oxygen and hydrogen submerged in a dilute sulfuric acid electrolyte solution, essentially reversing a water electrolysis reaction. Early development of fuel cells had a focus on use of coal to power fuel cells, but poisons formed by the gasification of the coal limited the fuel-cell usefulness and lifetime.[1] High-temperature solid oxide fuel cells (SOFCs) began with Nernst's 1899 demonstration of the still-used yttria-stabilized zirconia solid-state ionic conductor, but significant practical application was not realized.[2] The molten carbonate fuel cell (MCFC) utilizes a mixture of alkali metal carbonates retained in a solid ceramic porous matrix that become ionically conductive at elevated ($>600°C$) temperatures and was first studied for application as a direct coal fuel cell in the 1930s.[3] In 1933, Sir Francis Bacon began development of an alkaline-based oxygen-hydrogen fuel cell that achieved a short-term power density of 0.66 W/cm^2, high even for today's standards. However, little additional practical development of fuel cells occurred until the late 1950s, when the space race between the United States and the Soviet Union catalyzed development of fuel cells for auxiliary power applications. Low-temperature polymer electrolyte fuel cells (PEFCs) were first invented by William Grubb at General Electric in 1955 and generated power for NASA's *Gemini* space program. However, short operational lifetime and high catalyst loading contributed to a shift to alkaline fuel cells (AFCs) for the NASA *Apollo* program, and AFCs still serve as auxiliary power units (APUs) for the space shuttle orbiter.

After the early space-related application development of fuel cells went into relative abeyance until the 1980s. The phosphoric acid fuel cell (PAFC) became the first fuel cell system to reach commercialization in 1991. Although only produced in small quantities (twenty to forty 200 kW units per year) by United Technologies Company (UTC), UTC has installed and operated about 250 units similar to the 200 kW unit shown in Figure 1 in 19 countries worldwide for applications such as reserve power for the First National Bank of Omaha. As of 2002, these units have successfully logged over 5 million hours of operation with 95 percent fleet availability.[1]

Led by researchers at Los Alamos National Laboratory in the mid-1980s, resurgent interest in PEFCs was spawned through the development of an electrode assembly technique that enabled an order-of-magnitude reduction in noble-metal

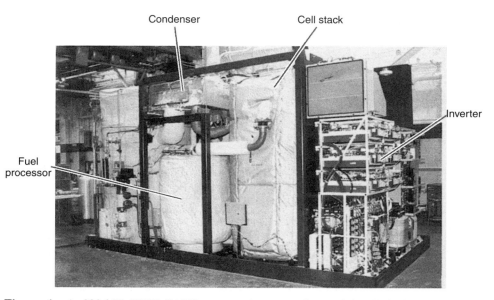

Figure 1 A 200-kW PC25 PAFC power plant manufactured by United Technologies Corporation. (Ref. 4.)

catalyst loading. This major breakthrough and ongoing environmental concerns, combined with availability of a non-hydrocarbon-based electrolyte with substantially greater longevity than those used in the Gemini program, has resurrected research and development of PEFCs for stationary, automotive, and portable power applications.

The area of research and development toward commercialization of high-temperature fuel cells, including MCFC and SOFC systems, has also grown considerably in the past two decades, with a bevy of demonstration units in operation and commercial sales of MCFC systems.

The science and technology of fuel cell engines are both fascinating and constantly evolving. This point is emphasized by Figure 2, which shows the registered fuel-cell-related patents in the United States, Canada, and United Kingdom since 1975. An acceleration of the patents granted in Japan and South Korea is also well underway, led by automotive manufacturer development. The nearly exponential growth in patents granted in this field is obvious and is not likely to wane in the near future.

Potential applications of fuel cells can be grouped into four main categories: (1) transportation, (2) portable power, (3) stationary power, and (4) niche applications. Although automotive fuel-cell applications have a great potential, they are also probably the least likely to be implemented on a large scale in the near future. The existing combustion engine technology market dominance will be difficult to usurp, considering its low comparative cost ($\sim$\$30/kW), high durability, high power density, suitability for rapid cold start, and high existing degree of optimization. Additionally, the recent success of high-efficiency hybrid

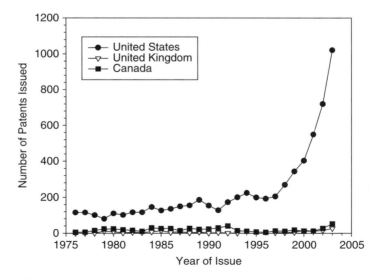

Figure 2   Timeline of worldwide patents in fuel cells. (Ref. 5.)

electric/combustion engine technology adds another rapidly evolving target fuel cells must match to compete.

Perhaps where fuel cells show the most promise for ubiquitous near-term implementation is in portable power applications, such as cell phones and laptop computers. Toshiba has recently developed a hand-held direct methanol fuel cell for portable power that is planned for sale in 2005. The 8.5 g direct methanol fuel cell (DMFC) is rated at 100 mW continuous power (up to 20 hours) and measures 22 mm × 56 mm × 4.5 mm with a maximum of 9.1 mm for the concentrated methanol fuel tank.[6] Passive portable fuel cells can potentially compete favorably with advanced lithium ion batteries in terms of gravimetric energy density of ∼120 to 160 Wh/kg and volumetric energy density of ∼230 to 270 Wh/L. Additionally, the cost of existing premium power battery systems is already on the same order as contemporary fuel cells, with additional development anticipated. With replaceable fuel cartridges, portable fuel-cell systems have the additional advantage of instant and remote rechargeability that can never be matched with secondary battery systems.

Stationary and distributed power applications include power units for homes or auxiliary and backup power generation units. Stationary applications (1 to 500 kW) are designed for nearly continuous use and therefore must have far greater lifetime than automotive units. Distributed power plants are designed for megawatt-level capacity, and some have been demonstrated to date. In particular, a 2-MW MCFC was recently demonstrated by Fuel Cell Energy in California.[7] A plot showing the estimated number of demonstration and commercial units in the stationary power category from 1986 to 2002 is given in Figure 3. Earlier growth corresponded mostly to PAFC units, although recently most additional units have been PEFCs. Not surprisingly, the exponential growth in the number of online units follows a similar qualitative trend to the available patents granted for various fuel-cell technologies shown in Figure 2. The early rise in stationary units in 1997 was primarily PAFC systems sold by United Technologies Center Fuel Cells. Data are estimated from the best available compilation available online at Ref. 8, and some manufacturers do not advertise prototype demonstrations, so that numbers are not exact. However, the trend is clear.

The fundamental advantages common to all fuel cell systems include the following:

1. A potential for a relatively high operating efficiency, scalable to all size power plants.
2. If hydrogen is used as fuel, greenhouse gas emissions are strictly a result of the production process of the fuel stock used.
3. No moving parts, with the significant exception of pumps, compressors, and blowers to drive fuel and oxidizer.

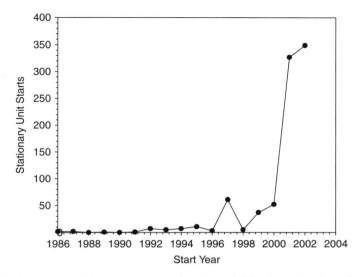

Figure 3 Estimated trend in the number of projects initiated to install stationary power sources since 1986. (Based on data from Ref. 9.)

4. Multiple choices of potential fuel feedstocks, from existing petroleum, natural gas, or coal reserves to renewable ethanol or biomass hydrogen production.

5. A nearly instantaneous and remote recharge capability compared to batteries.

Six technical limitations common to all fuel-cell systems must be overcome before successful implementation can occur:

1. Alternative materials and construction methods must be developed to reduce fuel-cell system cost to be competitive with the automotive combustion engine (~$30/kW) and stationary power systems (~$1,000/kW). The cost of the catalyst no longer dominates the price of most fuel-cell systems, although it is still significant. Manufacturing and mass production technology is now a key component to the commercial viability of fuel-cell systems.

2. Suitable reliability and durability must be achieved. The performance of every fuel cell gradually degrades with time, due to a variety of phenomena. The automotive fuel cell must withstand load cycling and freeze-thaw environmental swings with an acceptable level of degradation from the beginning-of-lifetime (BOL) performance over a lifetime of 5,000 hours (equivalent to 150,000 miles at 30 mph). A stationary fuel cell must withstand over 40,000 hours of steady operation under vastly changing external temperature conditions.

3. Suitable system power density and specific power must be achieved. The U.S. Department of Energy year 2010 targets for system power density and specific power are 650 W/kg and 650 W/L for automotive (50 kW) applications, 150 W/kg and 170 W/L for auxiliary (5-10 kW-peak) applications, and 100 W/kg and 100 W/L for portable (megawatt to 50 W) power systems.[10] Current systems fall well short of these targets.

4. Fuel storage and delivery technology must be advanced if pure hydrogen is to be used. The issue of hydrogen infrastructure (i.e., production, storage, and delivery), is not addressed herein but is nevertheless daunting in scope. This topic is addressed in detail in Ref. 11.

5. Fuel reformation technology must be advanced if a hydrocarbon fuel is to be used for hydrogen production.

6. Desired performance and longevity of system ancillary components must be achieved. New hardware (e.g., efficient transformers and high-volume blowers) will need to be developed to suit the needs of fuel-cell power systems.

The particular limitations and advantages of several different fuel-cell systems will be discussed in this chapter in greater detail.

2   BASIC OPERATING PRINCIPLES, EFFICIENCY, AND PERFORMANCE

Figure 4 shows a schematic of a generic fuel cell. Electrochemical reactions for the anode and cathode are shown for the most common fuel-cell types. Table 1 presents the various types of fuel cells, operating temperature, electrolyte material, and likely applications. The operating principle of a fuel cell is similar to a common battery, except that a fuel (hydrogen, methanol, or other) and oxidizer (commonly air or pure oxygen) are brought separately into the electrochemical reactor from an external source, whereas a battery has stored reactants. Referring to Figure 4, separate liquid-or gas-phase fuel and oxidizer streams enter through flow channels separated by the electrolyte/electrode assembly. Reactants are transported by diffusion and/or convection to the catalyzed electrode surfaces, where electrochemical reactions take place. Some fuel cells (alkaline and polymer electrolyte) have a porous (typical porosity $\sim$ 0.5–0.8) contact layer between the electrode and current-collecting reactant flow channels that functions to transport electrons and species to and from the electrode surface. In PEFCs, an electrically conductive carbon paper or cloth diffusion media (DM) layer (also called gas diffusion layer, or GDL) serves this purpose and covers the anode and cathode catalyst layer.

At the anode electrode, the electrochemical oxidation of the fuel produces electrons that flow through the bipolar plate (also called cell interconnect) to the external circuit, while the ions migrate through the electrolyte. The electrons in

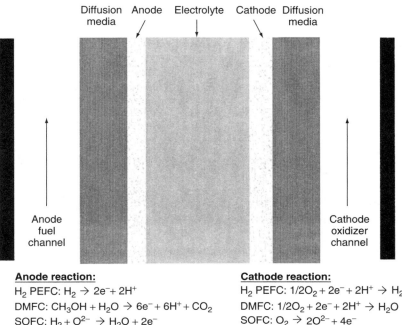

Diffusion media Anode Electrolyte Cathode Diffusion media

Anode fuel channel

Cathode oxidizer channel

Anode reaction:
H_2 PEFC: $H_2 \rightarrow 2e^- + 2H^+$
DMFC: $CH_3OH + H_2O \rightarrow 6e^- + 6H^+ + CO_2$
SOFC: $H_2 + O^{2-} \rightarrow H_2O + 2e^-$
AFC: $2H_2 + 4OH^- \rightarrow 4H_2O + 4e^-$
MCFC: $2H_2 + 2CO_3^{2-} \rightarrow 2H_2O + 2CO_2 + 4e^-$
PAFC: $H_2 \rightarrow 2e^- + 2H^+$

Cathode reaction:
H_2 PEFC: $1/2O_2 + 2e^- + 2H^+ \rightarrow H_2O$
DMFC: $1/2O_2 + 2e^- + 2H^+ \rightarrow H_2O$
SOFC: $O_2 \rightarrow 2O^{2-} + 4e^-$
AFC: $O_2 + 4e^- + 2H_2O \rightarrow 4OH^-$
MCFC: $O_2 + 2CO_2 + 4e^- \rightarrow 2CO_3^{2-}$
PAFC: $1/2O_2 + 2e^- + 2H^+ \rightarrow H_2O$

Figure 4 Schematic of a generic fuel cell.

the external circuit drive the load and return to the cathode catalyst where they recombine with the oxidizer in the cathodic oxidizer reduction reaction (ORR). The outputs of the fuel cell are thus threefold: (1) chemical products, (2) waste heat, and (3) electrical power.

A number of fuel-cell varieties have been developed to differing degrees, and the most basic nomenclature of fuel cells is related to the electrolyte utilized. For instance, a SOFC has a solid ceramic oxide electrolyte, and a PEFC has a flexible polymer electrolyte. Additional subclassification of fuel cells beyond the basic nomenclature can be assigned in terms of fuel used (e.g., hydrogen PEFC or direct methane SOFC) or the operating temperature range.

Each fuel-cell variant has particular advantages that engender use for particular applications. In general, low-temperature fuel cells (e.g., PEFCs, AFCs) have advantages in startup time and potential efficiency, while high-temperature fuel cells (e.g., SOFCs, MCFCs) have an advantage in raw materials (catalyst) cost and quality and ease of rejection of waste heat. Medium-temperature fuel cells (e.g., PAFCs) have some of the advantages of both high-and low-temperature classification. Ironically, a current trend in SOFC development is to enable lower

Table 1 Fuel-Cell Types, Descriptions, and Basic Data

Fuel-Cell Type	Electrolyte Material	Operating Temperature	Major Poison	Advantages	Disadvantages	Most Promising Applications
AFC	Solution of potassium hydroxide in water	60–250°C (modern AFCs <100°C)	CO_2	High efficiency, low oxygen reduction reaction losses	Must run on pure oxygen without CO_2 contaminant	Space applications with pure O_2/H_2 available
PAFC	Solution of phosphoric acid in porous silicon carbide matrix	160–220°C	Sulfur, high levels of CO	1–2% CO tolerant, good-quality waste heat, demonstrated, durability	Low-power density, expensive; platinum catalyst used; slow startup; loss of electrolyte	Premium stationary power
SOFC	Yttria (Y_2O_3) stabilized zirconia (ZrO_2)	600–1000°C	Sulfur	CO tolerant, fuel flexible, high-quality waste heat, inexpensive catalyst	Long start-up time, durability under thermal cycling, inactivity of electrolyte below ~600°C	Stationary power with cogeneration, continuous-power applications
MCFC	Molten alkali metal (Li/K or Li/Na) carbonates, in porous matrix	600–800°C	Sulfur	CO tolerant, fuel flexible, high-quality waste heat, inexpensive catalyst	Electrolyte dissolves cathode catalyst, extremely long start-up time, carbon dioxide must be injected to cathode, electrolyte maintenance	Stationary power with cogeneration, continuous-power applications
PEFC*	Flexible solid perfluorosulfonic acid polymer	30–100°C	CO, sulfur, metal ions, peroxide	Low-temperature operations, high efficiency, high H_2 power density, relatively rapid startup	Expensive catalyst, durability of components not yet sufficient, poor-quality waste heat, intolerance to CO, thermal and water management	Portable, automotive, and stationary applications

*Includes DMFC and direct alcohol fuel cells (DAFCs).

66

temperature ($<600°C$) operation, while a focus of current PEFC research is to operate at higher temperature ($>120°C$). Although the alkaline and phosphoric acid fuel cell had much research and development in the past and the MCFC is still under development, fuel-cell technologies under the most aggressive development are the PEFC and SOFC.

2.1 Description of a Fuel-Cell Stack

A single cell can be made to achieve whatever current and power are required simply by increasing the size of the active electrode area and reactant flow rates. However, the output voltage of a single fuel cell is always less than 1 V for realistic operating conditions, limited by the fundamental electrochemical potential of the reacting species involved. Therefore, for most applications and for compact design, a fuel-cell stack of several individual cells connected in series is utilized. Figure 5 is a schematic of a generic planar fuel-cell stack assembly and shows the flow of current through the system. For a stack in series, the total current is proportional to the active electrode area of the cells in the stack and is the same through all cells in the stack, and the total stack voltage is the sum of the individual cell voltages. For applications that benefit from higher voltage output, such as automotive stacks, it is typical to have over 200 fuel cells in a single stack.

Other components necessary for fuel-cell system operation include subsystems for oxidizer delivery, electronic control including voltage regulation, fuel storage and delivery, fuel recirculation/consumption, stack temperature control, and

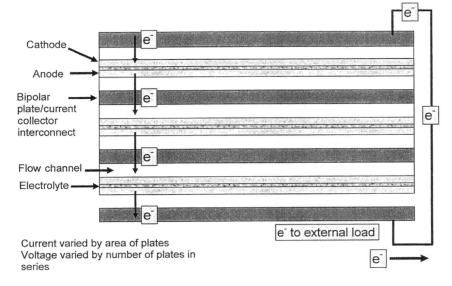

Figure 5 Schematic of the fuel-cell stack concept.

systems sensing of control parameters. For the PEFC, separate humidification systems are also needed to ensure optimal performance and stability. A battery is often used to start reactant pumps/blowers during start-up. In many fuel cells operating at high temperature, such as a SOFC or MCFC, a preheating system is needed to raise cell temperatures during start-up. This is typically accomplished with a combustion chamber that burns fuel and oxidizer gases. In all commercial fuel cells, provision must be made for effluent recovery. Fuel utilization efficiency is not 100 percent due to concentration polarization limitation on performance, so that unused fuel must be actively recycled, utilized, or converted prior to exhaust to the environment. Potential effluent management schemes include the use of recycling pumps, condensers (for liquid fuel), secondary burners, or catalytic converters.

2.2 Performance and Efficiency Characterization

The single-cell combination shown in Figure 4 provides a voltage dependent on operating conditions such as temperature, pressure, applied load, and fuel/oxidant flow rates. The thermal fuel-cell voltage (E_{th}) corresponds to a 100 percent efficient fuel cell and is shown as

$$E_{th} = \frac{\Delta H}{nF} \tag{1}$$

This is the total thermal voltage potential available if all chemical energy was converted to electrical potential. This is not possible, however, due to entropy change during reaction.

Consider a generalized global fuel cell reaction:

$$v_A A + v_B B \rightleftarrows v_C C + v_D D \tag{2}$$

The maximum possible electrochemical potential can be calculated from the Nernst equation as

$$E^0 = \frac{-\Delta G^0(T)}{nf} + \frac{RT}{nF} \ln\left(\frac{a_A^{v_A} a_B^{v_B}}{a_C^{v_C} a_D^{v_D}}\right) \tag{3}$$

where $\Delta G^0(T)$ is evaluated at the fuel-cell temperature and standard pressure and the pressure dependency is accounted for in the second term on the right. (See nomenclature for definitions of terms.) The activity of gas-phase reactants, besides water vapor, can be calculated from

$$a_i = \frac{y_i P}{P^0} \tag{4}$$

where P^0 is a standard pressure of 1 atm. Under less than fully saturated local conditions, the activity of water vapor can be shown to be the relative humidity:

$$a_{H_2O} = \frac{y_{H_2O} P}{P_{sat}(T)} = RH \tag{5}$$

where P_{sat} is the saturation pressure of water vapor at the fuel-cell temperature. For values of saturation pressure >1 atm ($\sim 100°C$), a standard pressure of 1 atm should be used for the water vapor activity.

Because a fuel cell directly converts chemical energy into electrical energy, the maximum theoretical efficiency is not bound by the Carnot cycle but is still not 100 percentage due to entropy change via reaction. The maximum thermodynamic efficiency is simply the ratio of the maximum achievable electrochemical potential to the maximum thermal potential:

$$\eta_{th} = \frac{E^0}{E_{th}} = \frac{\Delta G}{\Delta H} = \frac{\Delta H - T\Delta S}{\Delta H} = 1 - \frac{T\Delta S}{\Delta H} \qquad (6)$$

The actual operating thermodynamic (voltaic) efficiency of the fuel cell is the actual fuel cell voltage E_{fc} divided by E_{th}. The operating efficiency is really a ratio of the useful electrical output and heat output. If water is generated by the reaction, as is the case with most fuel-cell systems, efficiency and voltage values can be based on the low heating value (LHV, all water generated is in the gas phase) or the high heating value (HHV, all water generated is in the liquid phase). Representation with respect to the LHV accounts for the latent heat required to vaporize liquid water product, which is the natural state of the fuel-cell system.

Table 2 shows a selection of fuel-cell reactions and the calculated maximum theoretical efficiency at 298 K. Typical values for maximum efficiency at the open circuit calculated from equation (6) range from 60 to 90 percentage and vary with temperature according to sign of the net entropy change. That is, according to equation (6), the efficiency will decrease with temperature if the net entropy change is negative and increase with temperature if ΔS is positive (since $\Delta H\$$ is negative for an exothermic reaction). Based on the Le Chatelier principle, we can predict the qualitative trend in the functional relationship between maximum efficiency and temperature. If the global fuel-cell reaction ($v_A + v_B = v_C + v_D$) has more moles of gas-phase products than reactants (e.g., $v_C + v_D > v_A + v_B$), ΔS will be positive and η_{th} will increase with temperature. Liquid- or solid-phase

Table 2 Some Common Fuel-Cell Reactions and Maximum Theoretical Efficiency at 298 K

Fuel	Global Reaction	n (electrons per mole fuel)	Maximum η_{th} (HHV)
Hydrogen	$H_2 + \frac{1}{2}O_2 \rightarrow H_2O$	2	83
Methanol	$CH_3OH + 3 2O_2 \rightarrow CO_2 + 2H_2O$	6	97
Methane	$CH_4 + 2O_2 \rightarrow CO_2 + 2H_2O$	8	92
Formic acid	$HCOOH + \frac{1}{2}O_2 \rightarrow CO_2 + H_2O_1$	2	106
Carbon monoxide	$CO + \frac{1}{2}O_2 \rightarrow CO_2$	2	91
Carbon	$C_8 + \frac{1}{2}O_2 \rightarrow CO$	2	124

species have such low relative entropy compared to gas-phase species that they have negligible impact on ΔS. These fuels yield a theoretical maximum efficiency greater than 100 percentage! Physically, this means the reaction would absorb heat from the environment and convert the energy into a voltage potential. An example of this is a direct carbon fuel cell. Although use of ambient heat to generate power with an efficiency greater than 100 percentage seems like an amazing possibility, it is of course not realistic in practice due to various losses. If the number of gas-phase moles is the same between products and reactants, η_{th} is basically invariant with temperature and near 100 percentage, as in the methane-powered fuel cell with gas-phase water produced. Most fuel cells, including those with hydrogen fuel, have less product gas-phase moles compared to the reactants, and ΔS is negative. Thus, for these fuel cells η_{th} will *decrease* with operating temperature.

Figure 6 shows the calculated maximum thermodynamic efficiency of a hydrogen–air fuel cell with temperature compared to that of a heat engine. Since the maximum heat engine efficiency is the Carnot efficiency, it is an increasing function of temperature. Note that at a certain temperature above 600°C, the theoretical efficiency of the hydrogen–air fuel cell actually becomes *less* than that of a heat engine. In fact, high-efficiency combined-cycle gas turbines can now achieve power conversion efficiencies that rival high-temperature SOFCs. It is important to realize that fuel-cell systems are not inherently more efficient than heat engine alternatives. In practice, a 100 kW system operated by Dutch and Danish utilities has demonstrated an operating efficiency of 46 percent (LHV) over more than 3,700 hours of operation.[12] Combined fuel cell/bottoming cycle and cogeneration plants

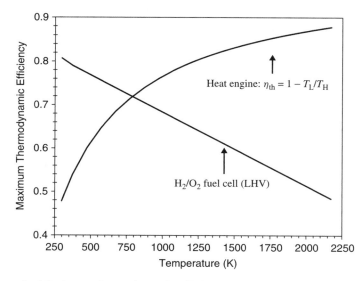

Figure 6 Maximum thermodynamic efficiency of a fuel cell and heat engine.

can achieve operational efficiencies as high as 80 percent with very low pollution. Another major advantage of fuel cells compared to heat engines is that efficiency is not a major function of device size, so that high-efficiency power for portable electronics can be realized, whereas small-scale heat engines have very low efficiencies due to heat transfer from high surface area–volume ratio. In terms of automotive applications, fuel cell hybrids and stand-alone systems operating with a variety of (but not all) fuel feedstocks have the potential for greater than double the equivalent mileage as conventional vehicles.[13]

Thermodynamic efficiency is not the entire picture, however, as the overall fuel-cell efficiency must consider the utilization of fuel and oxidizer. The appropriate mass flow rate of reactants into the fuel cell is determined by several factors related to the minimum requirement for electrochemical reaction, thermal management, and issues related to the particular type of fuel cell. However, the *minimum*-flow requirement is prescribed by the electrochemical reaction. An expression for the molar flow rate of species required for the electrochemical reaction can be shown from Faraday's law as[11]

$$\dot{n}_k = \frac{iA}{n_k F} \tag{7}$$

where i and A represent the current density and total electrode area, respectively, and n_k represents the electrons generated in the global electrode reaction per mole of reactant k. For fuel cells, the stoichiometric ratio or stoichiometry for an electrode reaction is defined as the ratio of reactant delivered to that required for the electrochemical reaction.

$$\text{Anode:} \quad \xi_a = \frac{\dot{n}_{\text{fuel,actual}}}{\dot{n}_{\text{fuel,required}}} = \frac{\dot{n}_{\text{fuel,actual}}}{iA/(n_{\text{fuel}}F)}$$

$$\text{Cathode:} \quad \xi_r = \frac{\dot{n}_{\text{ox,actual}}}{\dot{n}_{\text{ox,required}}} = \frac{\dot{n}_{\text{ox,actual}}}{iA/(n_{\text{ox}}F)} \tag{8}$$

The stoichiometry can be different for each electrode and must be greater than unity. This is due to the fact that zero concentration near the fuel-cell exit will result in zero voltage from equation (3). Since the current collectors are electrically conductive, a large potential difference cannot exist and cell performance will decrease to zero if reactant concentration goes to zero. As a result of this requirement, fuel and oxidizer utilization efficiency is never 100 percent, and some system to recycle or consume the effluent fuel from the anode is typically required to avoid releasing unused fuel to the environment. Thus, the overall operating efficiency of a fuel cell can be written as the product of the voltaic and Faradaic (fuel utilization) efficiencies:

$$\eta_{\text{fc}} = \eta_{\text{th}} \frac{1}{\zeta_a} \frac{1}{\zeta_c} \tag{9}$$

2.3 Polarization Curve

Figure 7 is an illustration of a typical polarization curve for a fuel cell with negative ΔS, such as the hydrogen–air fuel cell, showing five regions labeled I-V. The polarization curve, which represents the cell voltage–current relationship, is the standard figure of merit for evaluation of fuel-cell performance. Also shown in Figure 7 are the regions of electrical and heat generation. Since the thermodynamically available power not converted to electrical power is converted to heat, the relationship between current and efficiency can be clearly seen by comparing the relative magnitude of the voltage potential converted to waste heat and to electrical power. Region V is the departure from the maximum thermal voltage, caused by entropy generation. In practice, the open-circuit voltage (OCV) achieved is somewhat less than that calculated from the Nernst equation. Region IV represents this departure from the calculated maximum open-circuit voltage. This loss can be very significant and for PEFCs is due to undesired species crossover through the thin-film electrolyte and resulting mixed potential at the electrodes. For other fuel cells, there can be some loss generated by internal currents from electron leakage through the electrolyte. This is especially a challenge in SOFCs. Beyond the departure from the theoretical open-circuit potential, there are three major classifications of losses that result in a drop of the fuel-cell voltage potential, shown in Figure 7: (1) activation (kinetic) polarization (region I), (2) ohmic polarization (region II), and (3) concentration polarization (region III). It should be noted that voltage loss,

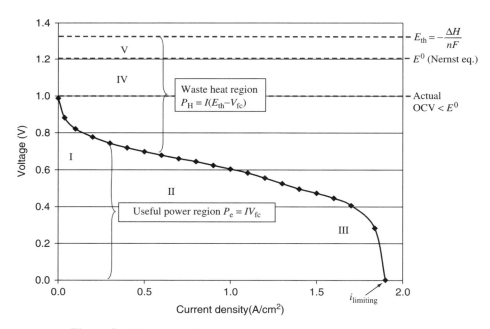

Figure 7 Illustration of a typical polarization curve for a fuel cell.

polarization, and overpotential are all interchangeable and refer to a voltage loss. The operating voltage of a fuel cell can be represented as the departure from ideal voltage caused by these polarizations:

$$E_{\text{fc}} = E^0 - \eta_{\text{a,a}} - |\eta_{\text{a,c}}| - \eta_r - \eta_{\text{m,a}} - |\eta_{\text{m,c}}| \tag{10}$$

where E^0 is the theoretical Nernst open-circuit potential of the cell [equation (3)] and the activation overpotential at the anode and cathode are represented by $\eta_{\text{a,a}}$ and $\eta_{\text{a,a}}$, respectively. The ohmic (resistive) polarization is shown as η_r. The concentration overpotential at the anode and cathode are represented as $\eta_{\text{m,a}}$ and $\eta_{\text{m,c}}$, respectively. Cathode polarization losses are negative relative to the standard hydrogen electrode, so the absolute value is taken. Additional losses in region IV of Figure 7 attributed to species crossover or internal currents can also be added to equation (10) as needed. Activation and concentration polarizations occur at both anode and cathode, while the resistive polarization represents ohmic losses throughout the fuel cell and thus includes ionic, electronic, and contact resistances between all fuel-cell components carrying current. It is important to note that the regions on the polarization curve of dominance kinetic, ohmic, or mass-transfer polarizations are not discrete. That is, all modes of loss contribute throughout the entire current density range, and there is no discrete *ohmic loss* region where other polarizations are not also contributing to total deviation below the Nernst potential. Although the activation overpotential dominates in the low-current region, it still contributes to the cell losses at higher current densities where ohmic or concentration polarization dominates. Thus, each region is not discrete, and all types of losses contribute throughout the operating current regime.

Activation polarization, which dominates losses at low current density, is the voltage overpotential required to overcome the activation energy of the electrochemical reaction on the catalytic surface and is commonly represented by a Butler–Volmer equation at each electrode:[15]

$$i_{\text{tc}} = i_0 \left[\exp\left(\frac{\alpha_{\text{a}} F n}{RT} \eta \right) - \exp\left(\frac{\alpha_{\text{c}} F n}{RT} \eta \right) \right] \tag{11}$$

where α_{a} and α_{c} are the charge-transfer coefficients for the anode and cathode, respectively. The fraction of the electrical overpotential (η) resulting in a change of the rate of reaction for the cathodic branch of this electrode is shown as α_{c}. Obviously, $\alpha_{\text{a}} + \alpha_{\text{c}} = 1$. Here, n is the number of exchange electrons involved in the elementary electrode reaction, which is typically different from the n used in equations (3) and (7). The exchange current density i_0 represents the activity of the electrode for a particular reaction. In hydrogen PEFCs, the anode i_0 for hydrogen oxidation is so high relative to the cathode i_0 for oxygen reduction that the anode contribution to this polarization is often neglected for pure hydrogen fuel. On the contrary, if neat hydrogen is not used, significant activation polarization losses at both electrodes are typical (e.g., the DMFC). It appears from equation (11) that activation polarization should increase with temperature. However, i_0 is a highly nonlinear function of the kinetic rate constant of reaction and the local reactant concentration and can be modeled with an Arrhenius form as

$$i_0 = i_o^0 \exp^{E_A(RT)} \left(\frac{C_{ox}}{C_{ref}} \right)^{\gamma} \left(\frac{C_f}{C_{ref}} \right)^{\upsilon} \tag{12}$$

Thus, i_0 is an exponentially increasing function of temperature, and the net effect of increasing temperature is to decrease activation polarization. For this reason, high-temperature fuel cells such as SOFCs or MCFCs typically have very low activation polarization and can use less exotic catalyst materials. Accordingly, the effect of an increase in electrode temperature is to decrease the voltage drop within the activation polarization region shown in Figure 7. For various fuel-cell systems, however, the operating temperature range is dictated by the electrolyte and materials properties, so that temperature cannot be arbitrarily increased to reduce activation losses.

At increased current densities, a primarily linear region is evident on the polarization curve. In this region, reduction in voltage is dominated by internal ohmic losses (η_r) through the fuel cell that can be represented as

$$\eta_r = i A \left(\sum_{k=1}^{n} r_k \right) \tag{13}$$

where each r_k value is the area-specific resistance of individual cell components, including the ionic resistance of the electrolyte, and the electric resistance of bipolar plates, cell interconnects, contact resistance between mating parts, and any other cell components. With proper design and assembly, ohmic polarization is typically dominated by electrolyte conductivity for all fuel-cell types.

At very high current densities, mass-transport limitation of fuel or oxidizer to the corresponding electrode causes a sharp decline in the output voltage. This is referred to as concentration polarization. This region of the polarization curve is really a combined mass-transport/kinetic related phenomenon, as the surface reactant concentration is functionally related to the exchange current density, as shown in equation (12). In general, reactant transport to the electrode is limited to some value depending on, for example, operating conditions, porosity and tortuosity of the porous media, and input stoichiometry. If this limiting mass-transport rate is approached by the consumption rate of reactant [equation (7)], the surface concentration of reactant will approach zero, and, from equation (3), the fuel-cell voltage will also approach zero. The Damköler number (Da) is a dimensionless parameter that is the ratio of the characteristic electrochemical reaction rate to the rate of mass transport to the reaction surface. In the limiting case of infinite kinetics (high Damköler number), one can derive an expression for η_m as

$$\eta_m = -\frac{RT}{nF} \ln\left(1 - \frac{i}{i_l}\right) = -B \ln\left(1 - \frac{i}{i_l}\right) \tag{14}$$

where i_l is the limiting current density and represents the maximum current produced when the surface concentration of reactant is reduced to zero at the reaction site. The limiting current density (i_l) can be determined by equating the reactant consumption rate to the mass transport rate to the surface, which in itself can be a complex calculation. The strict application of equation (14) results in a predicted voltage drop-off that is much more abrupt than actually observed. To accommodate a more gradual slope, an empirical coefficient B is often used to fit the model to the experiment.[15] Concentration polarization can also be incorporated into the exchange current density and kinetic losses, as in equation (12).

2.4 Heat Management

Although PEFC systems can achieve a high relative operating efficiency, the inefficiencies manifest as dissipative thermal losses. At the cell level, if waste heat is not properly managed, accelerated performance degradation or catastrophic failure can occur. The total waste heat rate can be calculated as

$$P_{\text{waste}} = kI(E_{\text{th}} - E_{\text{fc}}) \tag{15}$$

where k is the number of cells in series in the stack.

This heat generation can be broken down into components and shown as[16]

$$\begin{aligned}
q^n = {} & -i_{\text{fc}}\left(\frac{\Delta H}{nF} - \frac{\Delta G}{nF} - \eta_{\text{a,a}} - |\eta_{\text{a,c}}| - i_{\text{fc}}\sum_{k=1}^{n} r_k\right) \\
& + i_{\text{fc}}^2 \sum_{k=1}^{n} r_k - i_{\text{fc}}\left(-\eta_{\text{a,a}} - |\eta_{\text{a,c}}| + \frac{T\Delta S}{nF}\right)
\end{aligned} \tag{16}$$

The first term on the right-hand side of equation (16) is Joule heating and is thus an $i^2 r$ relationship. The second and third terms of equation (16) represent the heat flux generated by activation polarization in the anode and cathode catalyst layers. This assumes the concentration dependence on exchange current density is included in the activation polarization terms. The third term in equation (16) is a linearly varying function of current density and represents the total Peltier heat generated via entropy change by reaction. The functional relationship derived in equation (16) is shown in Figure 8, a plot of the heat generation via Peltier, Joule, and kinetic heating as a function of current density for a typical PEFC.[17] The ionic conductivity for the electrolyte was chosen to be 0.1 S/cm, based on the assumption of a fully humidified membrane in contact with vapor-phase water,[18] and other parameters were chosen as typical values. This plot should be viewed

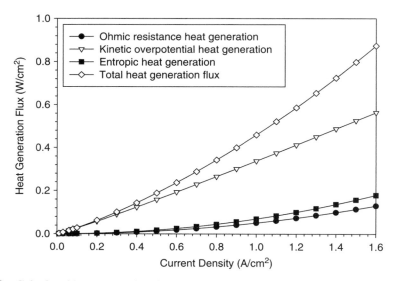

Figure 8 Calculated heat generation from activation, ohmic, and entropic sources as a function of current for a typical PEFC.

only as a guide to the qualitative behavior of the heat generation with current density, as each fuel cell and different operating conditions will have much different distributions. For example, the SOFC typically has quite low activation polarization generated heat, due to the high operating temperatures. Note that an assumption of ohmic heating dominance is not always accurate, and entropic heat generation can be quite significant and cannot be ignored. For higher-temperature fuel cells with low activation overpotential, the heat generation is dominated by ohmic and entropic terms.

2.5 Degradation

The lifetime of a fuel cell is expected to compete with existing power systems it would replace. As a result, the automotive fuel cell must withstand load cycling and freeze–thaw environmental swings with minimal degradation over a lifetime of 5,000 hours. A stationary fuel cell, meanwhile, must withstand more than 40,000 hours of steady operation with minimal downtime. The fuel-cell environment is especially conducive to degradation, since a voltage potential difference exists that can promote undesired reaction, and some fuel cells operate at high temperature or have corrosive electrolytes. Transient load cycling between high- and low-power points has also been shown to accelerate degradation, so that steady-state degradation rates may not be truly representative for transient systems. Many different modes of physicochemical degradation are known to exist, including these four:

1. *Catalyst or electrolyte poisoning and degradation.* Since the catalyst and electrolyte control the reaction and ohmic polarization, any poisoning or other degradation of these components will adversely affect cell performance. Some minor species in air and re-formed gas product, such as carbon monoxide, will foul a platinum catalyst operating at low temperatures ($<150°C$). As a result, PEFCs are extremely susceptible to CO poisoning. Some surface absorption of species can be reversible, including adsorbed CO. Sulfur is also a major contaminant that will greatly reduce performance in most fuel cells in extremely low ($<$ppb) concentrations. Many other low-level impurities can greatly harm fuel-cell performance; for example, CO_2 will degrade the electrolyte in AFCs. Given that the electrolyte is an ion conductor, when unintended ions are present in the fuel-cell system by corrosion or other impurities, the electrolyte will absorb these impurities, which can alter the ionic conductivity of the media. In SOFCs, carbonaceous residue from internal performance can foul the anode catalyst.

2. *Electrolyte loss.* In some cases, electrolyte material is lost through a variety of physicochemical mechanisms. For polymer electrolyte fuel cells, the polymer itself can degrade physically and chemically, particularly from peroxide radical attack.[19] This results in loss of mass and conductivity in the electrolyte and possible catastrophic pinhole formation. For liquid electrolyte systems such as the AFC, PAFC, and MCFC, the finite vapor pressure of the liquid phase results in a steady but predictable loss of electrolyte through the reactant flow streams, which must be replenished with regularity or performance will suffer.

3. *Morphology changes or loss in catalyst layer or other components.* For all fuel cells, the catalyst layer electrochemical active surface area (ECSA) is a determining factor in overall power density, and nanosize catalysts and supports are present in a complex, three-dimensional electrode structure designed to simultaneously optimize electron, ion, and mass transfer. As a result, any morphological changes can result in reduced performance. Commonly observed phenomena include catalyst sintering, dissolution and migration, catalyst oxidation, supporting material oxidation (e.g., carbon corrosion for carbon-supported catalysts), and Oswald ripening.[20] These effects are most often irreversible. Other components can also be chemically or physically altered, such as the porosity distribution or hydrophobicity of the GDL in a PEFC.

4. *Corrosion of other components.* Oxidation of other components such as the current collector can become a major loss in fuel cells over time. This is especially relevant in high-temperature MCFCs and SOFCs, where the corrosion process is accelerated by the high temperature. Chromium used in SOFCs in stainless-steel interconnects is believed to cause cathode

degradation. Low-temperature PEFCs can also suffer losses from current collector corrosion, so proper coatings with high electronic conductivity must be used.[21]

2.6 Hydrogen PEFC

The hydrogen (H_2) PEFC is seen by many as the most viable alternative to heat engines and battery replacement for automotive, stationary, and portable power applications. The H_2 PEFC is fueled either by pure hydrogen or from a diluted hydrogen mixture generated from a hydrocarbon re-formation process. An H_2 PEFC fuel-cell stack power density of greater than 1.0 kW/L is typical.

2.7 H_2 PEFC Performance

Hydrogen PEFCs operate at $60°$ to $100°C$. The anode and cathode catalyst is commonly ~ 2 nm platinum or platinum-ruthenium powder supported on significantly larger-size carbon particles with a total (anode and cathode) platinum loading of ~ 0.4 mg/cm^2. This represents a major breakthrough in required catalyst loading from the 28 mg/cm^2 of the original 1960s H_2 PEFC. As a result, the catalyst is no longer the dominating factor in fuel-cell cost, although it is still higher than needed to reach long-term goals of another 20-fold reduction in loading or elimination of precious metals.[22] The state-of-the-art H_2 PEFC can reach nearly 0.7 V at 1 A/cm^2 under pressurized conditions at $80°C$. It is always desirable to operate at high voltages because of increased efficiency and reduced flow requirements. However, power density typically peaks below 0.6 V, and heat generation can cause accelerated degradation, so there is a size trade-off for high-voltage operation. Typical anode and cathode stoichiometry requirements are low, with typical values of 2 or less for the cathode and even lower values for the anode.

2.8 Technical Issues in H_2 PEFC

The H_2 PEFC has many technical issues that complicate performance and control. Besides issues of manufacturing, ancillary system components, cost, and market acceptance, the main remaining technical challenges for the fuel cell include (1) water and heat management, (2) durability, and (3) freeze–thaw cycling capability.

Water and Heat Management
For PEFCs, waste heat affects the water distribution by increasing temperature and thus the local equilibrium saturation pressure of the gases. At a typical PEFC operating temperature of $80°C$ and atmospheric pressure, each $1°C$ change in temperature results in an approximately 5 percent change in equilibrium saturation pressure.[23] Thus, the thermal and water management and control

are inexorably coupled at the individual cell and stack level, and even small variations in temperature can dramatically affect optimal humidity, locations of condensation/vaporization, membrane longevity, and a host of other phenomena. Due to heat generation, relatively high temperature gradients up to 10°C between the electrolyte and current collector can occur.[24] Analysis has shown that the through-plane thermal conductivities of carbon cloth gas diffusion media and Nafion electrolyte material are approximately 0.15 and 0.1 W/mK, respectively.[25,26] The main barrier to the heat transport is through the GDL, which acts as a thermal insulator to limit conductive heat transfer. Once through the GDL, the majority of heat transfer is typically through the landings and not to the reactant in the flow channels.[17]

For high-power (>kW) fuel-cell stacks, waste heat must be properly managed with cooling channels, which take up space and require parasitic pumping losses. The choice of coolant is based on the necessary properties of high specific heat, nonconductive, noncorrosive, sufficient boiling/freezing points for operation in all environments, and low viscosity. Laboratory systems typically use deionized water, although practical systems exposed to the environment must use a lower freezing point nonconductive solution.

Water management and humidification are major issues in H_2 PEFC performance. The most common electrolyte used in PEFCs is a perfluorosulfonic acid–polytetrafluoroethylene (PTFE) copolymer in acid (H^+) form, known by the commercial name Nafion (E. I. du Pont de Nemours). Nafion electrolyte conductivity is primarily a function of water content and temperature, shown as[18]

$$\sigma_e = 100 \exp\left[1268\left(\frac{1}{303} - \frac{1}{T(\text{K})}\right)\right](0.005139\lambda - 0.00326) \quad (\mho \cdot \text{m}^{-1})$$

(17)

$$\lambda = 0.043 + 17.18a - 39.85a^2 + 36.0a^3 \text{ for } 0 < a \leq 1$$
where
$$\lambda = 14 + 14(a - 1)$$
$$a = \text{water activity} = y_{H_2O}P/P_{sat}(T) = \text{RH}$$

A plot of Nafion conductivity as a function of humidity, based on equation (17), is given in Figure 9. It is obvious that ionic conductivity is severely depressed without sufficient water. Alternatively, excessive water at the cathode can cause flooding, that is, liquid water accumulation at the cathode surface that prevents oxygen access to the reaction sites. Flooding is most likely near the cathode exit under high-current-density, high-humidification, low-temperature, and low-flow-rate conditions. However, the term *flooding* has been rather nebulously applied in the literature to date, representing a general performance loss resulting from liquid water accumulation blocking reactant transport to the electrode. There are actually six discrete regions that can suffer flooding losses in the PEFC: the anode and cathode catalyst layers, the anode and cathode gas diffusion layers, and the anode and cathode flow channels. Figure 10 is a radiograph

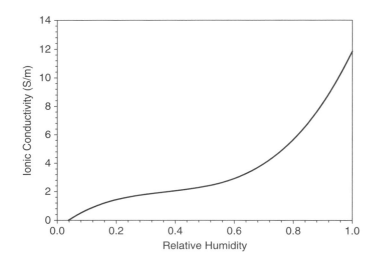

Figure 9 Electrolyte conductivity of Nafion as a function of relative humidity at 80°C.

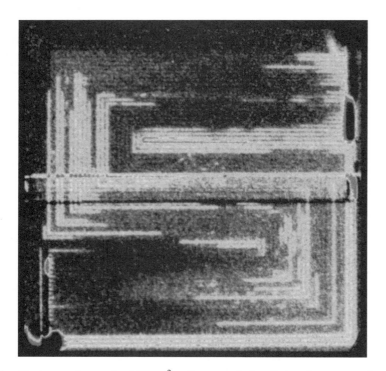

Figure 10 Neutron radiograph of $50\,cm^2$ active area fuel cell showing severe channel-level flooding at low current density for a H_2 PEFC. (From Ref. 29.)

image of a PEFC under severely flooded conditions with significant diffusion media and channel-level flooding. In this image, the water accumulation was determined to be primarily in the anode flow channels. This phenomenon can occur for low-flow-rate, low-power, and high-fuel-utilization conditions, which indicates that channel level flooding is not solely a cathode issue.

Depending on operating conditions, flow field design, and material properties, a membrane can have a highly nonhomogeneous water (and therefore ionic conductivity) distribution. The membrane beneath a long channel may be dried by hot inlet flow ideally saturated near the middle of the cell and experiencing flooding near the exit. It is difficult in practice to maintain an ideal water distribution throughout the length of the cell. Thus, water transport is an especially important issue in PEFC design. Even a slight reduction in ohmic losses through advanced materials, thinner electrolytes, or optimal temperature/water distribution can significantly improve fuel-cell performance and power density.

Figure 11 shows a schematic of the water-transport and generation modes in the PEFC. At the cathode surface, the oxygen reduction reaction results in water production proportional to the current density via Faraday's law. Water transport through the electrolyte occurs by diffusion, electroosmotic drag, and hydraulic permeation from a pressure difference across the anode and cathode. Diffusion through the electrolyte can be represented using Fick's law, and appropriate expressions relating to diffusion coefficients for Nafion can be found.[27] An electroosmotic drag coefficient (λ_{drag}) of 1 to 5 H_2O/H^+ of Nafion membranes has been shown for a fully hydrated Nafion 117 membrane.[28] The drag coefficient was shown to be a nearly linearly increasing function of temperature from $20°$ to $120°C$. Thus, the water delivered to the cathode by this transport mode can be 2 to 10 times greater than that generated by reaction. Hydraulic permeation

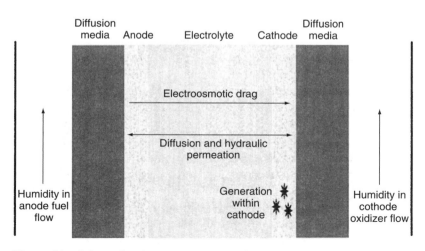

Figure 11 Schematic of water transport and generation modes in the PEFC.

of water through the membrane under gas-phase pressure difference between the anode and cathode is usually small for H_2 PEFCs. However, capillary pressure differences can result in a net flux of water via this mode of transport.[30] Combining the different forms of water transport through the membrane, the bulk molar water transport and creation at the cathode can be shown as

$$\dot{n}_{H_2O} = -DA\frac{\Delta C_{c,a}}{\Delta x} + \frac{iA}{F}(\lambda_{drag} + 0.5) - \dot{n}_{\text{pressure driven,a,c}} \qquad (18)$$

where the last term on the right represents the flow of water by imbalanced pressure forces, which can be further represented by capillary- and gas-phase pressure terms. Since the reduction reaction and electroosmotic drag result in water transport to and generation at the cathode surface, the flux of water by diffusion can be either to or from the anode surface, depending on local flow and humidity conditions. The flux by pressure difference is a function of tailorable material and operating properties and can flow toward either electrode, depending on the desire of the designer.

To achieve a proper water balance, many technologies have been employed, including tailored temperature or pressure gradients to absorb the net water generated. Flow field design also has a profound effect on the local liquid water distribution and is too detailed a topic to discuss in this chapter. The ambient relative humidity also should affect the system water balance; however, this effect is typically quite small due to the vast difference between typical ambient temperature and operating temperature $P_{g,sat}$ values. For example, even fully saturated inlet flow drawn from an ambient air source at $20°C$ ($P_{g,sat} = 2.338\,kPa$) contains only 5 percent of the water required for saturation in a fuel cell operating at $80°C$ ($P_{g,sat} = 47.39\,kPa$).

Durability and Freeze–Thaw Cycling

For PEFCs, durability and freeze–thaw cycling capability are major issues. Durability issues in PEFCs have been discussed for fuel cells in general. For PEFCs, carbon support corrosion, morphological changes, susceptibility to chemical poisoning (especially carbon monoxide), catalyst loss, and electrolyte degradation all contribute to an operational degradation rate that is on the order of microvolts per hour under ideal steady-state conditions, which is still too high.[31] Load cycling is known to initiate even greater degradation rate for a variety of reasons, so that steady-state laboratory testing typically underestimates true performance loss. Under nonoptimal higher temperature or low-humidity conditions, longevity is even less. Additionally, further system size and weight reductions are needed for automotive packaging requirements.

For automotive and stationary applications, PEFCs must withstand vastly changing environments as low as $-40°C$. Because the electrolyte contains water, freezing results in some ice formation and volume change, which can easily result in damage to the electrode structure and interfacial contact between the various layer structures. To compete with existing combustion-based technology, fuel-cell stacks must achieve over 100 cold starts at $-20°C$ with drivable power in under 5 seconds.

2.9 Direct Methanol Fuel Cell

The liquid-fed DMFC is seen as the most viable alternative to lithium ion batteries in portable applications because DMFC systems require less ancillary equipment and can therefore be more simplified compared to an H_2 PEFC. While both H_2 PEFCs and DMFCs are strictly PEFCs (same electrolyte), the DMFC feeds a liquid solution of methanol and water to the anode as fuel. The additional complexities of the low-temperature internal re-formation prevent the DMFC from obtaining the same level of fuel-cell power density as the H_2 PEFC. For the DMFC, both anode and cathode activation polarizations are significant and are the same order of magnitude. However, reduced performance compared to the H_2 PEFC is tolerable in light of other advantages of the DMFC:

1. Because anode flow is mostly liquid (gaseous CO_2 is a product of methanol oxidation), there is no need for a separate cooling or humidification subsystem.

2. Liquid fuel used in the anode results in lower parasitic pumping requirements compared to gas flow. In fact, an emerging class of passive DMFC designs operate without any external parasitic losses, instead relying on natural forces such as capillary action, buoyancy, and diffusion to deliver reactants.

3. The highly dense liquid fuel stored at ambient pressure eliminates problems with fuel storage volume. With highly concentrated methanol as fuel (>10 M), passive DMFC system power densities can compare favorably to advanced Li ion batteries.

2.10 Technical Issues of the DMFC

Four main technical issues affecting performance remain: (1) water management, (2) methanol crossover, (3) managing two-phase transport in the anode, and (4) high activation polarization losses and catalyst loading. While significant progress has been made by various groups to determine alternative catalysts, total catalyst loading is still on the order of 10 mg/cm^2. Typically a platinum–ruthenium catalyst is utilized on the anode to reduce polarization losses from CO intermediate poisoning, and a Pt catalyst is used on the cathode.[32]

External humidification is not needed in the DMFC, due to the liquid anode solution, but prevention of cathode flooding is critical to ensure adequate performance. Flooding is more of a concern for DMFCs than H_2 PEFCs because of constant diffusion of liquid water to the cathode. To prevent flooding, cathode airflow must be adequate to remove water at the rate that it arrives and is produced at the cathode surface. Assuming thermodynamic equilibrium, the minimum stoichiometry required to prevent liquid water accumulation in the limit of zero water diffusion through the membrane can be shown by equating equations (1) and (2) as[33]

$$\xi_{x,\min} = \frac{2.94/P_{x,\text{sat}}}{P_1 - P_{g,\text{sat}}} + 1 \qquad (19)$$

The factor of 1 in equation (19) is a result of the consumption of oxygen in the cathode by an electrochemical oxygen reduction reaction. For most cases, the minimum cathode stoichiometry for a DMFC is determined by flooding avoidance rather than electrochemical requirements. Therefore, optimal cathode stoichiometries are significantly greater than unity. It should be noted that equation (19) is purely gas phase and therefore does not allow for water removal in the liquid phase, as droplets, a capillary stream out of the cathode or to the anode, or even entrained as a mist in the gas flow. Some recirculation of the water is needed, or, in a totally passive system design, the hydrophobicity of the GDL and catalyst can be tailored to pump water back to the anode via capillary action.

Another critical issue in the DMFC is methanol crossover from the anode to the cathode. This is a result of diffusion, electroosmotic drag, and permeation from pressure gradients. Therefore, an expression for the methanol crossover through the membrane can be written similar to equation (18) with different transport properties. When crossover occurs, the mixed potential caused by the anodic reaction on the cathode electrode reduces cell output and the true stoichiometry of the cathode flow.

Of the three modes of methanol crossover, diffusion (estimated as $10^{-5.4163-999.778/T}$ m^2/s)[34] is dominant under normal conditions, especially at higher temperatures. Since the driving potential for oxidation is so high at the cathode, the methanol that crosses over is almost completely oxidized to CO_2, which sets up a sustained maximum activity gradient in methanol concentration across the electrolyte. The electroosmotic drag coefficient of methanol (estimated as 0.16 CH_2OH/H^+,[35] or $2.5y$, where y is the mole fraction of CH_3OH in solution[36]) is relatively weak owing to the nonpolar nature of the molecule. To prevent crossover so that more concentrated (and thus compact) solutions of methanol can be utilized as fuel, various diffusion barriers have been developed. That is, a porous filter in the GDL or separating the fuel from the fuel channel can be used to separate concentrated methanol solution from the membrane electrode assembly (MEA), greatly reducing crossover through the electrolyte.[37] An earlier solution was to use a thicker electrolyte to reduce methanol crossover, but the concomitant loss in performance via increased ohmic losses through the electrolyte was unsatisfactory.

Several other transport-related issues are important to DMFC performance. The anode side is a two-phase system primarily consisting of methanol solution and product CO_2. The methanol must diffuse to the catalyst, while the reaction-generated CO_2 must diffuse outward from the catalyst. At high current densities, CO_2 can become a large volume fraction ($>90\%$) in the anode flow field. Carbon monoxide removal from the catalyst sites is critical to ensure adequate methanol oxidation. Other disadvantages of the DMFC are related to use of methanol. Methanol is toxic, can spread rapidly into groundwater, has a colorless flame, and is more corrosive than gasoline.

3 SOLID OXIDE FUEL CELL

The SOFC and MCFC represent high-temperature fuel-cell systems. The current operating temperature of most SOFC systems is around $800°$ to $1000°C$, although new technology has demonstrated $600°C$ operation, where vastly simplified system sealing and materials solutions are feasible. High electrolyte temperature is required to ensure adequate ionic conductivity (of O^{2-}) in the solid-phase ceramic electrolyte and reduces activation polarization so much that cell losses are typically dominated by internal cell ohmic resistance through the electrolyte. Typical SOFC open-circuit cell voltages are around 1 V, very close to the theoretical maximum, and operating current densities vary greatly depending on design. While the theoretical maximum efficiency of the SOFC is less than the H_2 PEFC because of increased temperature, activation polarization is extremely low, and operating efficiencies as high as 60 percent have been attained for a 220 kW cogeneration system.[38]

There has been much recent development in the United States on SOFC systems, incubated by the Department of Energy Solid State Energy Conversion Alliance (SECA) program. The 10-year goal of the SECA program is to develop kilowatt-size SOFC APU units at $400/kW with rated performance achievable over the lifetime of the application with less than 0.1 percent loss per 500 hr operation by 2021.

The solid-state, high-temperature ($600–1000°C$) SOFC system eliminates many of the technical challenges of the PEFC while suffering unique limitations. The SOFC power density varies greatly depending on cell design but can achieve above $400 \, mW/cm^2$ for some designs. In general, a SOFC system is well suited for applications where a high operating temperature and a longer start-up transient are not limitations, or where conventional fuel feedstocks are desired.

There are three main advantages of the SOFC system:

1. High operating temperature greatly reduces activation polarization and eliminates the need for expensive catalysts. This also provides a tolerance to a variety of fuel stocks and enables internal reformation of complex fuels.
2. High-quality waste heat, enabling a potential for high overall system efficiencies ($\sim80\%$) utilizing a bottoming or cogeneration cycle.[39]
3. Tolerance to CO, which is a major poison to Pt-based low-temperature PEFCs.

3.1   Technical Issues of SOFC

Besides manufacturing and economic issues, which are beyond the scope of this chapter, the main technical limitations of the SOFC include operating temperature, long start-up time, durability, and cell-sealing problems resulting from mismatched thermal expansion of materials. For additional details, an excellent text for SOFC was written by Minh and Takahashi.[40]

The high operating temperature of the SOFC requires long start-up time to avoid damage due to nonmatched thermal expansion properties of materials. Another temperature-related limitation is that no current generation is possible until a critical temperature is reached in the solid-state electrolyte, where oxygen ionic conductivity of the electrolyte becomes non-negligible. Commonly used electrolyte conductivity is nearly zero until around 650°C,[15] although low-temperature SOFC operation at 500°C using doped ceria (CeO_2) ceramic electrolytes has shown feasibility.[41] In many SOFC designs, a combustor is utilized to burn fuel and oxidizer effluent to preheat the cell to light-off temperature and hasten start-up and provide a source of heat for cogeneration. In addition, the combustor effectively eliminates unwanted hydrogen or CO, which is especially high during start-up when fuel-cell performance is low. Additionally, electrolyte, electrode, and current collector materials must have matched thermal expansion properties to avoid internal stress concentrations and damage during both manufacture and operation.

The desire for lower temperature operation of the SOFC is ironically opposite to the PEFC, where higher operating temperature is desired to simplify water management and CO poisoning issues. Lower temperature ($\sim$400–500°C) operation would enable rapid start-up, use of common metallic compounds for cell interconnects, reduce thermal stresses, reduce the rate of some modes of degradation, and increase reliability and reduced manufacturing costs. Despite the technical challenges, the SOFC system is a good potential match for many applications, including stationary cogeneration plants and auxiliary power.

Durability of SOFCs is not solely related to thermal mismatch issues. The electrodes suffer a strong poisoning effect from sulfur (in the ppb range), requiring the use of sulfur-free fuels. Additionally, anode oxidation of nickel catalysts can decrease performance and will do so rapidly if the SOFC is operated below 0.5 V. As SOFC reduced operating temperature targets are achieved and use of inexpensive metallic interconnects become feasible, accelerated degradation from metallic interaction is possible. Metals utilizing chromium (e.g., stainless steels) have shown limited lifetimes in SOFCs and can degrade the cathode catalyst through various physicochemical pathways.[42]

3.2   Performance and Materials

In the SOFC system, yttria- (Y_2O_3-) stabilized zirconia (ZrO_2) is most often used as the electrolyte. In contrast to PEFCs, O^{2-} ions are passed *from the cathode to the anode* via oxygen vacancies in the electrolyte instead of H^+ ions *from the anode to the cathode*. Other cell components such as interconnects and bipolar plates are typically doped ceramic, cermet, or metallic compounds.

There are four different basic designs for the SOFC system: planar, sealless tubular, monolithic, and segmented cell-in-series designs. Two of the designs, planar and tubular, are the most promising for continued development. The other designs have been comparatively limited in development to date. The planar configuration looks geometrically similar to the generic fuel-cell shown in Figure 4. The three layer anode–electrolyte–cathode structure can be an anode-, cathode-, or electrolyte-supported design, meaning the structural support is provided by a thicker layer of one of the structures (anode, cathode, or electrolyte supported). Since excessive ionic and concentration losses result from electrolyte- and cathode-supported structures, respectively, many designs utilize an anode-supported structure, although ribbed supports or cathode-supported designs are utilized in some cases.

For the planar design, the flow channel material structure is used as support for the electrolyte, and a stacking arrangement is employed. Although this design is simple to manufacture, one of the major limitations is difficulty in sealing the flow fields at the edges of the fuel cell. Sealing is a key issue in planar SOFC design because it is difficult to maintain system integrity over the large thermal variation and reducing/oxidizing environment over many start-up and shut-down load cycles. Compressive, glass, cermet, glass–ceramic, and hybrid seals have been used with varied success for this purpose.

The second major design is the sealless tubular concept pioneered by Westinghouse (now Siemens-Westinghouse) in 1980. A schematic of the general design concept is shown in Figure 12. Air is injected axially down the center of the fuel cell, which provides preheating of the air to operation temperatures before exposure to the cathode. The oxidizer is provided at adequate flow rates to ensure negligible concentration polarization at the cathode exit, to maintain desired cell temperature, and to provide adequate oxidizer for effluent combustion with unused fuel. The major advantage of the tubular configuration is that the difficult high-temperature seals needed for the planar SOFC design are eliminated. Tubular designs have been tested in 100 kW atmospheric pressure and 250-kW pressurized demonstration systems with little performance degradation with time (less than 0.1% per 1,000 hr) and efficiencies of 46 percent and 57 percent (LHV), respectively.[38]

One drawback of this type of tubular design is the more complex and limited range of cell fabrication methods. Another drawback is high internal ohmic losses relative to the planar design due to the relatively long in-plane path that electrons must travel along the electrodes to and from the cell interconnect. Some of these additional ionic transport losses have been reduced by use of a flattened tubular SOFC design with internal ribs for current flow, called the high-power-density (HPD) design by Siemens-Westinghouse and shown schematically in Figure 13.[43]

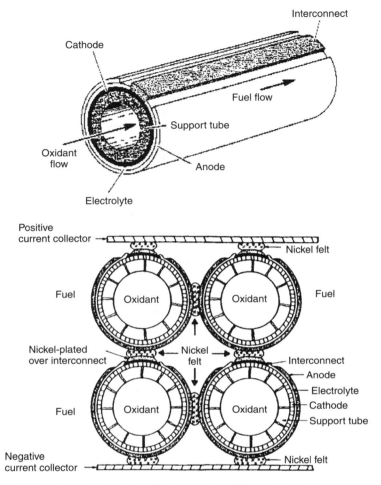

Figure 12 Schematic of sealles tubular SOFC design. (From Ref. 40.)

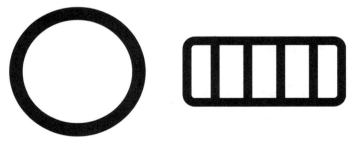

Standard tubular design concept   Flattened tubular design concept

Figure 13 End-view schematic of conventional tubular and flattened high-power-density SOFC concepts for reduced ionic transport losses.

This design can also experience significant losses due to limited oxygen transport through the porous ($\sim$35% porosity) structural support tube used to provide rigidity to the assembly. The internal tube can also be used as the anode, reducing these losses through the higher diffusivity of hydrogen.

The monolithic and segmented cell-in-series designs are less developed, although demonstration units have been constructed and operated. A schematic of the monolithic cell design is shown in Figure 14. In the early 1980s, the corrugated monolithic design was developed based on the advantage of HPD compared to other designs. The HPD of the monolithic design is a result of the high active area exposed per volume and the short ionic paths through the electrolyte, electrodes, and interconnects. The primary disadvantage of the monolithic SOFC design, preventing its continued development, is the complex manufacturing process required to build the corrugated system.

The segmented cell-in-series design has been successfully built and demonstrated in two configurations: the bell-and-spigot and banded configurations shown schematically in Figure 15. The bell-and-spigot configuration uses stacked segments with increased electrolyte thickness for support. Ohmic losses are high because electron motion is along the plane of the electrodes in both designs, requiring short individual segment lengths ($\sim$1–2 cm). The banded configuration avoids some of the high ohmic losses of the bell-and-spigot configuration with a thinner electrolyte but suffers from the increased mass-transport losses associated with the porous support structure used. The main advantage of the segmented cell design is a higher operating efficiency than larger area single-electrode

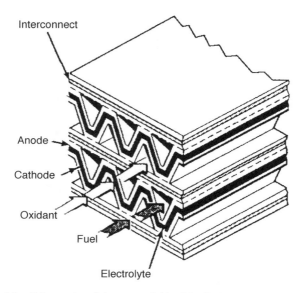

Figure 14 Schematic of the monolithic SOFC design. (From Ref. 40.)

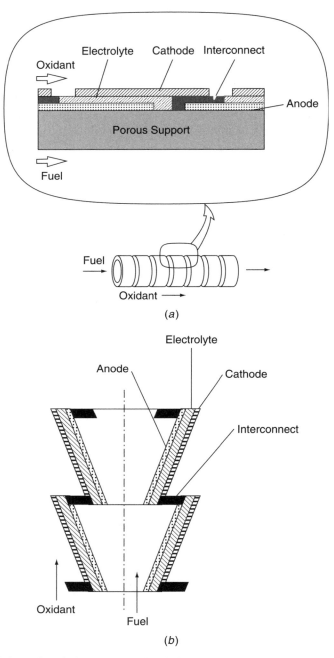

Figure 15   Schematic of the segmented cell-in-series design: (*a*) banded and (*b*) bell-and-spigot configuration. (From Ref. 40.)

configurations. The primary disadvantages limiting development of the segmented cell designs include the necessity for many high-temperature gas-tight seals, relatively high internal ohmic losses, and requirement for manufacture of many segments for adequate power output. Other cell designs, such as radial configurations and more recently microtubular designs, have been developed and demonstrated to date.

4   OTHER FUEL CELLS

Many other fuel-cell varieties and configurations, too numerous to enumerate here, have been developed to some degree. The most developed to date have been the phosphoric acid, alkaline, and molten carbonate fuel cells, all three of which have had some applied and commercial success.

4.1   Alkaline Fuel Cells

Alkaline fuel cells utilize a solution of potassium hydroxide in water as an alkaline, mobile (liquid) electrolyte. Alkaline fuel cells were originally developed as an APU for space applications by the Soviet Union and the United States in the 1950s and served in the *Apollo* program as well as on the current space shuttle orbiter. Figure 16 shows AFCs for installation on *Apollo* mission service modules in 1964.[44] Alkaline fuel cells were chosen for space applications for their high efficiency and robust operation. Both circulated and static electrolyte designs have been utilized. The AFC operates at around $60°$ to $250°$C with greatly varied electrode design and operating pressure. More modern designs tend to operate at the lower range of temperature and pressure. The primary advantages of the AFC are the cheaper cost of materials and electrolyte and high operating efficiency (60 percent demonstrated for space applications) due to use of an alkaline electrolyte. For alkaline electrolytes, ORR kinetics are much more efficient than acid-based electrolytes (e.g., PEFCs, PAFCs) enabling high relative operating efficiencies. Since space applications typically utilize pure oxygen and hydrogen for chemical propulsion, the AFC was well suited. However, the electrolyte suffers an intolerance to even small fractions of CO_2 found in air, which reacts to form potassium carbonate (K_2CO_3) in the electrolyte, gravely reducing performance over time. For terrestrial applications, CO_2 poisoning has limited the lifetime of AFCs systems to well below that required for commercial application, and filtration of CO_2 is too expensive for practical use. Due to this limitation, relatively little commercial development of the AFC beyond space applications has been realized.

4.2   Molten Carbonate Fuel Cells

Currently MCFCs are commercially available from several companies, including a 250 kW unit from Fuel Cell Energy in the United States and several other

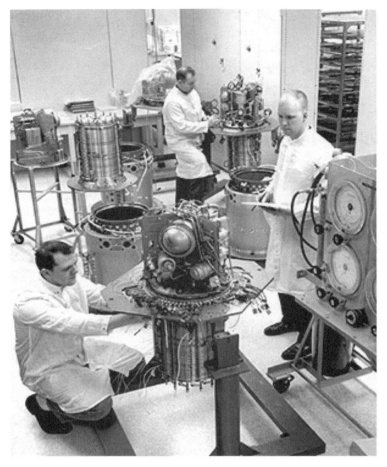

Figure 16   Pratt & Whitney technicians assemble alkali fuel cells for Apollo service modules, 1964. (From Ref. 43.)

companies in Japan. Some megawatt-size demonstration units are installed world-wide, based on natural gas or coal-based fuel sources which can be internally re-formed within the anode of the MCFC. The MCFCs operate at high temperature (600–700°C) with a molten mixture of alkali metal carbonates (e.g., lithium and potassium) or lithium and sodium carbonates retained in a porous ceramic matrix. In the MCFC, CO_3^{2-} ions generated at the cathode migrate to the anode oxidation reaction. The MCFC design is similar to a PAFC, in that both have liquid electrolytes maintained at precise levels within a porous ceramic matrix and electrode structure by a delicate balance of gas-phase and capillary pressure forces. A major advantage of the MCFC compared to the PAFC is the lack of precious-metal catalysts, which greatly reduces the system raw material costs. Original development on the MCFC was mainly funded by the U.S.

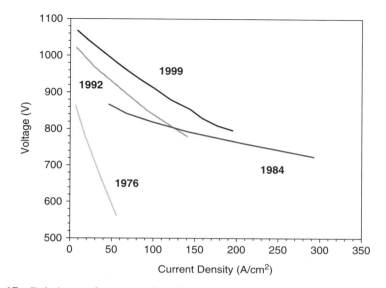

Figure 17 Relative performance of MCFCs at 1 atm pressure. (Adapted from Ref. 15.)

Army in the 1950s and 1960s, and significant advances of this liquid electrolyte high-temperature fuel-cell alternative to the SOFC were made.[45] The U.S. Army desired operation of power sources from logistic fuel, thus requiring high temperatures with internal fuel re-formation that can be provided by the MCFC. Development waned somewhat after this early development, but advances have continued and initial commercialization has been achieved. The steadily increasing performance of MCFCs throughout years of development is illustrated in Figure 17. The main advantages of MCFCs include the following:

- The MCFC can consume CO as a fuel and generates water at the anode, thus making it ideal for internal re-formation of complex fuels.
- As with the SOFC, high-quality waste heat is produced for bottom-cycle or cogeneration applications.
- Non-noble-metal catalysts are used, typically nickel–chromium or nickel–aluminum on the anode and lithiated nickel oxide on the cathode.

The main disadvantages of MCFCs include the following:

- Versus the SOFC (*the other high-temperature fuel cell*), the MCFC has a highly corrosive electrolyte that, coupled with the high operating temperature, accelerates corrosion and limits longevity of cell components, especially the cathode catalyst. The cathode catalyst (nickel oxide) has a significant dissolution rate into molten carbonate electrolyte.
- Extremely long start-up time (the MCFC is generally suitable only for continuous power operation) and nonconductive electrolyte at low temperatures.

- Carbon dioxide must be injected into the cathode to maintain ionic conductivity. This can be accomplished with recycling from the anode effluent or injection of combustion product but complicates the system design.
- Electrolyte maintenance is an engineering technical challenge. The liquid electrolyte interface between the electrodes is maintained by a complex force balance involving gas-phase and electrolyte-liquid capillary pressure between the anode and cathode. Significant spillage of the electrolyte into the cathode can lead to catalyst dissolution but is difficult to eliminate.
- Vapor pressure of the electrolyte is nonnegligible and leads to loss of electrolyte through reactant flows.

4.3 Phosphoric Acid Fuel Cells

The PAFC was originally developed for commercial application in the 1960s. The PAFC has an acidic, mobile (liquid) electrolyte of phosphoric acid contained by a porous silicon carbide ceramic matrix and operates at around $160°$ to $220°C$. Like the MCFC, the electrolyte is bound by capillary and gas pressure forces between porous electrode structures. The PAFC is in many ways similar to the PEFC, except the acid-based electrolyte is in liquid form and the operating temperature is slightly higher. Over two hundred 200 kW commercial PAFC units were developed and sold by International Fuel Cells (now United Technologies Research Fuel Cells), and many are still in operation. However, ubiquitous commercial application has not been achieved, primarily due to the high cost (approximately \$4,500/kW), about five times greater than cost targets for conventional stationary applications. The main advantages of the PAFC include the following:

- The high operating temperature provides better waste heat than the PEFC and allows operation with 1 percent to 2 percent CO in the fuel stream, which is much better for use of reformed fuel compared to the PEFC, which cannot tolerate more than 10 ppm of CO without significant performance loss.[46]
- The acid electrolyte does not need water for conductivity, making water management very simple compared to the PEFC. Only the product water from the cathode reduction reaction needs to be removed, a relatively simple task at elevated temperatures.
- The demonstrated long life and commercial success for premium stationary power of the PAFC.

The main disadvantages of the PAFC include the following:

- The PAFC is a bulky, heavy system compared to the PEFC. Area specific power is less than for the PEFC (0.2–0.3 W/cm^2).[47]
- The use of platinum catalyst with nearly the same loading as PEFCs.

- The liquid electrolyte has finite vapor pressure, resulting in continual loss of electrolyte in the vapor phase and a continual need for replenishment or recirculation. Modern PAFC design includes cooling and condensation zones to mitigate this loss.

- The relatively long warm-up time until the electrolyte is conductive at ~160°C (although much less than for the MCFC or SOFC).

4.4 Other Alternatives

Many other fuel-cell systems exist, and new versions are constantly being developed. Most are simply existing fuel-cell systems with a new fuel. For example, PEFCs based on a direct alcohol solution offer alternatives to DMFCs for portable power application and include those based on formic acid,[48] dimethyl ether,[49] ethylene glycol, dimethyl oxalate, and others.[50] A completely different concept is the biological or microbial fuel cell (MFC). In the MFC, electricity is generated by anerobic oxidation of organic material by bacteria. The catalytic activity and transport of protons is accomplished using biological enzymes or exogenous mediators.[51,52] Although relative performance is very low, on the order of 1 to 100 mW/m^2, the potential for generating some power or simply power-neutral decomposition and treatment of organic waste matter such as sewage water is potentially quite significant to society.

Based on the continued growth and expansion of fuel-cell science, it is evident that, despite lingering technical challenges, continued growth and development of a variety of fuel cell systems will evolve toward implementation in many, but certainly not all, potential applications. In some cases, development of existing or new power sources or existing technical barriers will ultimately doom ubiquitous application of fuel cells, while some applications are likely to enjoy commercial success.

NOMENCLATURE

a	activity coefficient of species, unitless
A	area term, cm^2
C	molar concentration, mol/cm^3
D	diffusion coefficient, cm^2/s
E	voltage, V
E_a	activation energy of electrochemical reaction, J/mol
F	Faraday constant, charge on 1 mol of electrons, 96,487 As/mol electron
G	Gibbs free energy, kJ/kg
H	enthalpy, kJ/kg
I	current, A

i	current density, A/cm^2
i_0	exchange current density, A/cm^2
i_l	mass-limited current density, A/cm^2
M	solution molarity, mol/L
n	electrons per mole oxidized or reduced, e$^-$/mol
n.	molar flow rate, mol/s
P	pressure, Pa, and power, W
q''	heat flux, W/cm^2
r	area specific resistance, Ω/cm^2
R	universal gas constant, 8.314 J/mol·K
RH	relative humidity, unitless
S	entropy, kJ/kg·K
T	temperature, K
y	mole fraction, unitless

Greek Letters

α	charge transfer coefficient
$\Delta(x)$	change of parameter x
γ	reaction order for elementary oxidizer reaction
η	polarization, V
λ	electroosmotic drag coefficient, mol/mol H$^+$, and water saturation in electrolyte
σ	conductivity, $\Omega \cdot m^{-1}$
ν	stoichiometric coefficient of balanced equation; reaction order for elementary fuel reaction
ξ	stoichiometric flow ratio

Superscripts

0	standard conditions

Subscripts

a	activation or anode
c	cathode
e	electrolyte
fc	fuel cell
f	fuel
H$_2$O	water
i	species i
k	individual cell components; number of cells in a stack
m	mass transport

ox	oxidizer
r	resistive
ref	reference
sat	at saturation conditions
th	thermal

REFERENCES

1. M. L. Perry and T. F. Fuller, "A Historical Perspective of Fuel Cell Technology in the 20th Century," *J. Electrochem. Soc.*, **149**(7), pp. S59–S67 (2002).

2. W.Nernst, "Über die elektrolytische Leitung fester Körper bei sehr hohen Temperaturen," *Z. Elektrochem.*, No. 6, 41–43 (1899).

3. E. Baur and J. Tobler, *Z. Elektrochem. Angew. Phys. Chem.*, **39**, 180 (1933).

4. J. M. King and B. McDonald, "Experience with 200 kW PC25 Fuel Cell Power Plant," in *Handbook of Fuel Cells—Fundamentals, Technology and Applications*, Vol. 3, W. Vielstich, A. Lamm, and H. A. Gasteiger (eds.), John Wiley, West Chichester, UK, 2003, pp. 832–843.

5. http://www.upsto.gov; www.patent.gov.uk; www.patent.gov.uk.

6. Toshiba Press Release of June 24, 2004: http://www.toshiba.com/taec/press/dmfc__04__222.shtml.

7. Y. Mugikura, "Stack Materials and Stack Design," in *Handbook of Fuel Cells—Fundamentals, Technology and Applications*, Vol. 4, W. Vielstich, A. Lamm, and H. A. Gasteiger (eds.), Wiley, Chichester, UK, 2003, pp. 907–920.

8. http://www.fuelcells.org.

9. http://www.fuelcells.org.

10. United States Department of Energy 2003 Multi-Year Research, Development and Demonstration Plan for Fuel Cells: http://www.eere.energy.gov/hydrogenandfuelcells/mypp/pdfs/3.4__fuelcells.pdf.

11. D. Sperling, and J. S. Cannon, *The Hydrogen Energy Transition: Moving Toward the Post Petroleum Age in Transportation*, New York, Elsevier, 2004.

12. O. Yamamoto, "Solid Oxide Fuel Cells: Fundamental Aspects and Prospects," *Electrochem. Acta*, **45**, 2423–2435 (2000).

13. M. Wang, "Fuel Choices for Fuel Cell Vehicles: Well-to-Wheels Energy and Emission Impacts," *J. Power Sources*, **112**, 307–312 (2002).

14. A. J. Bard and L. R. Faulkner, *Electrochemical Methods*, John Wiley, New York, 1980.

15. J. Larmine and A. Dicks, *Fuel Cell Systems Explained*, 2nd ed., Wiley, Chichester, UK, 2003.

16. A. B. LaConti, M. Hamdan, and R. C. McDonald, "Mechanisms of Membrane Degradation," in *Handbook of Fuel Cells—Fundamentals, Technology and Applications*, Vol. 3, W. Vielstich, A. Lamm, and H. A. Gasteiger (eds.), Wiley, Chichester, UK, 2003, pp. 647–662.

17. D. J. Burford, *Real-Time Electrolyte Temperature Measurement in an Operating Polymer Electrolyte Membrane Fuel Cell*, Thesis, Penn State University, University Park, PA, 2004.

18. T. E. Springer, T. A. Zawodzinski and S. Gottesfeld, *J. Electrochem. Soc.*, **136**, 2334 (1991).

19. J. Divisek, "Low Temperature Fuel Cells," in *Handbook of Fuel Cells—Fundamentals, Technology and Applications*, Vol. 1, W. Vielstich, A. Lamm, and H. A. Gasteiger (eds.), Wiley, Chichester, UK, 2003, pp. 99–114.

20. D. P. Wilkinson and J. St-Pierre, "Durability," in *Handbook of Fuel Cells—Fundamentals, Technology and Applications*, Vol. 3, W. Vielstich, A. Lamm, and H. A. Gasteiger (eds.), Wiley, Chichester, UK, 2003, pp. 611–626.

21. D. A. Shores and G. A. DeLuga, "Basic Materials and Corrosion Issues," in *Handbook of Fuel Cells—Fundamentals, Technology and Applications*, Vol. 2, W. Vielstich, A. Lamm, and H. A. Gasteiger (eds.), Wiley, Chichester, UK, 2003, pp. 273–285.

22. http://www.eere.energy.gov/hydrogenandfuelcells/fuelcells/fc_parts.html#catalyst.

23. M. J. Moran and H. N. Shapiro, *Fundamentals of Engineering Thermodynamics*, 3rd ed., John Wiley, New York, 1995.

24. D. Burford, T. Davis, and M. M. Mench, "*In situ* Temperature Distribution Measurement in an Operating Polymer Electrolyte Fuel Cell," *Proceedings of 2003 ASME International Mechanical Engineering Congress and Exposition*, November 15–21, 2003.

25. D. Burford, S. He, and M. M. Mench, "Heat Transport and Temperature Distribution in PEFCs," *Proceedings of 2004 ASME International Mechanical Engineering Congress and Exposition*, November 13–19, 2004.

26. P. J. S. Vie and S. Kjelstrup, "Thermal Conductivities from Temperature Profiles in the Polymer Electrolyte Fuel Cell," *Electrochim. Acta*, **49**, 1069–1077 (2004).

27. S. Motupally, A. J. Becker, and J. W. Weidner, "Diffusion of Water in Nafion 115 Membranes," *J. Electrochem. Soc.*, **147**(9), 3171–3177 (2000).

28. X. Ren and S. Gottesfeld, "Electro-Osmotic Drag of Water in Poly(perfluorosulfonic acid) Membranes," *J. Electrochem. Soc.*, **148**(1), A87–A93 (2001).

29. N. Pekula, K. Heller, P. A. Chuang, A. Turhan, M. M. Mench, J. S. Brenizer, and K. Ünlü, "Study of Water Distribution and Transport in a Polymer Electrolyte Fuel Cell Using Neutron Imaging," *Nuclear Instruments and Methods in Physics Research Section: Accelerators Spectrometers Detectors and Associated Equipment*, Vol. 542, Issues 1–3, No. 21, pp. 134–141, 2005.

30. U. Pasaogullari and C. Y. Wang, "Two-Phase Transport and the Role of Micro-Porous Layer in Polymer Electrolyte Fuel Cells," *Electrochim. Acta*, **49**, 4359–4369 (2004).

31. S. Cleghorn, J. Kolde, and W. Liu, "Catalyst Coated Composite Membranes," in *Handbook of Fuel Cells—Fundamentals, Technology and Applications*, Vol. 3, W. Vielstich, A. Lamm, and H. A. Gasteiger (eds.), Wiley, Chichester, UK, 2003, pp. 566–575.

32. J. Müller, G. Frank, K. Colbow, and D. Wilkinson, "Transport/Kinetic Limitations and Efficiency Losses," in *Handbook of Fuel Cells—Fundamentals, Technology and Applications*, Vol. 4, W. Vielstich, A. Lamm, and H. A. Gasteiger (eds.), John Wiley, Chichester, UK, 2003, pp. 847–855.

33. M. M. Mench and C. Y. Wang, "An *in Situ* Method for Determination of Current Distribution in PEM Fuel Cells Applied to a Direct Methanol Fuel Cell," *J. Electrochem. Soc.*, **150**(1), A79–A85 (2003).

34. C. L. Yaws, *Handbook of Transport Property Data: Viscosity, Thermal Conductivity, and Diffusion Coefficients of Liquids and Gases*, Gulf, Houston, TX, 1995.

35. J. Cruickshank and K. Scott, "The Degree and Effect of Methanol Crossover in the Direct Methanol Fuel Cell," *J. Power Sources*, **70**, 40–47 (1998).

36. X. Ren, T. E. Springer, T. A. Zawodzinski, and S. Gottesfeld, *J. Electrochem. Soc.*, **147**, 466 (2000).

37. Y. Pan, *Integrated Modeling and Experimentation of Electrochemical Power Systems*, Thesis, Penn State University, University Park, PA, 2004.

38. R. F. Service, "New Tigers in the Fuel Cell Tank," *Science*, 288, 1955–1957 (2000).

39. S. C. Singhal, "Science and Technology of Solid-Oxide Fuel Cells," *MRS Bulletin*, March 2000, pp. 16–21.

40. N. Minh and T. Takahashi, *Science and Technology of Ceramic Fuel Cells*, Elsevier, New York, 1995.

41. R. Doshi, V. L. Richards, J. D. Carter, X. Wang, and M. Krumpelt, "Development of Solid Oxide Fuel Cells that Operate at 500°C," *J. Electrochem. Soc.*, 146, 1273–1278 (1999).

42. Y. Matsuzaki and I. Yasuda, "Dependence of SOFC Cathode Degradation by Chromium-Containing Alloy on Compositions of Electrodes," *J. Electrochem. Soc.*, **148**(2), A126–A131 (2001).

43. D. Stöver, H. P. Buchkremer, and J. P. P. Huijsmans, "MEA/Cell Preparation Methods: Europe/ USA," in *Handbook of Fuel Cells—Fundamentals, Technology and Applications*, Vol. 4, W. Vielstich, A. Lamm, and H. A. Gasteiger (eds.), John Wiley, Chichester, UK, 2003, pp. 1015–1031.

44. Image 059-016 from the Science Service Historical Image Collection, The Smithsonian Institution: http://americanhistory.si.edu/csr/fuelcells/alk/alk3.htm.

45. B. S. Baker (ed.), *Hydrocarbon Fuel Cell Technology*, Academic, New York, 1965.

46. R. D. Breault, "Stack Materials and Stack Design," in *Handbook of Fuel Cells—Fundamentals, Technology and Applications*, Vol. 4, W. Vielstich, A. Lamm, and H. A. Gasteiger (eds.), Wiley, Chichester, UK, 2003, pp. 797–810.

47. *Fuel Cell Handbook*, 5th ed., EG&G Services Parsen, Inc., Science Applications International Corporation, 2000.

48. M. Zhao, C. Rice, R. I. Masel, P. Waszczuk, and A. Wieckowski, "Kinetic Study of Electro-oxidation of Formic Acid on Spontaneously-Deposited Pt/Pd Nanoparticles—CO Tolerant Fuel Cell Chemistry," *J. Electrochem. Soc.*, **151**(1), A131–A136 (2004).

49. M. M. Mench, H. M. Chance, and C. Y. Wang "Dimethyl Ether Polymer Electrolyte Fuel Cells for Portable Applications," *J. Electrochem. Soc.*, **151**, A144–A150 (2004).

50. E. Peled, T. Duvdevani, A. Aharon, and A. Melman, "New Fuels as Alternatives to Methanol for Direct Oxidation Fuel Cells," *Electrochem. Solid-State Lett.*, **4**(4), A38–A41 (2001).

51. H. Liu, R. Narayanan, and B. Logan, "Production of Electricity during Wastewater Treatment Using a Single Chamber Microbial Fuel Cell," *Environ. Sci. Technol.* **38**, 2281–2285 (2004).

52. T. Chen, S. Calabrese Barton, G. Binyamin, Z. Gao, Y. Zhang, H.-H. Kim, and A. Heller, "A Miniature Biofuel Cell," *J. Am. Chem. Soc.*, **123**, 8630–8631 (2001).

CHAPTER 4

GEOTHERMAL RESOURCES AND TECHNOLOGY: AN INTRODUCTION

Peter D. Blair**
National Academy of Sciences
Washington, DC

1   INTRODUCTION   101

2   GEOTHERMAL
    RESOURCES   102
    2.1   The United States Geothermal
          Resource Base   103
    2.2   Hydrothermal Resources 104
    2.3   Hot Dry Rock and Magma
          Resources   106
    2.4   Geopressured Resources 106

3   GEOTHERMAL ENERGY
    CONVERSION   107
    3.1   Direct Uses of Geothermal
          Energy   109
    3.2   Electric Power Generation 109
    3.3   Geothermal Heat Pumps   116

1   INTRODUCTION

Geothermal energy is heat from Earth's interior. Nearly all of geothermal energy refers to heat derived from the Earth's molten core. Some of what is often referred to as geothermal heat derives from solar heating of the surface of the Earth, although it amounts to a very small fraction of the energy derived from the Earth's core. For centuries, geothermal energy was apparent only through anomalies in the Earth's crust that permit the heat from Earth's molten core to venture close to the surface. Volcanoes, geysers, fumaroles, and hot springs are the most visible surface manifestations of these anomalies.

Earth's core temperature is estimated by most geologists to be around $5,000°$ to $7,000°$C. For reference, that is nearly as hot as the surface of the sun (although, substantially cooler than the sun's interior). And although the Earth's core is cooling, it is doing so very slowly in a geologic sense, since the thermal conductivity of rock is very low and, further, the heat being radiated from Earth is being

---

**Peter D. Blair, PhD, is Executive Director of the Division on Engineering and Physical Sciences of the National Academy of Sciences (NAS) in Washington, DC. The views expressed in the chapter, however, are his own and not necessarily those of the NAS.

Table 1   Worldwide Geothermal Power Generation (2002) (Ref. 1)

Country	Installed Capacity (mWe)	Electricity Generation (millions kWh/yr)
United States	2,850	15,900
Philippines	1,848	8,260
Mexico	743	5,730
Italy	742	5,470
Japan	530	3,350
Indonesia	528	3,980
New Zealand	364	2,940
El Salvador	105	550
Nicaragua	70	276
Costa Rica	65	470
Iceland	51	346
Kenya	45	390
China	32	100
Turkey	21	90
Russia	11	30
Azores	8	42
Guadalupe	4	21
Taiwan	3	—
Argentina	0.7	6
Australia	0.4	3
Thailand	0.3	2
TOTALS	7,953	47,967

substantially offset by radioactive decay and solar radiation. Some scientists estimate that over the past three billion years, Earth may have cooled several hundred degrees.

Geothermal energy has been used for centuries, where it is accessible, for aquaculture, greenhouses, industrial process heat, and space heating. It was first used for production of electricity in 1904 in Lardarello, Tuscany, Italy, with the first commercial geothermal power plant (250 kWe) developed there in 1913. Since then geothermal energy has been used for electric power production all over the world, but most significantly in the United States, the Philippines, Mexico, Italy, Japan, Indonesia, and New Zealand. Table 1 lists the current levels of geothermal electric power generation installed worldwide.

2   GEOTHERMAL RESOURCES

Geothermal resources are traditionally divided into the following three basic categories or types that are defined and described later in more detail:

1. *Hydrothermal convection systems*, which include both vapor-dominated and liquid dominated systems

Table 2 Geothermal Resource Classification

Resource Type	Temperature Characteristics
Hydrothermal convection resources (heat carried upward from depth by convection of water or steam)	
a. Vapor dominated	−240°C
b. Liquid(hot-water) dominated	
1. High temperature	150–350°C
2. Intermediate temperature	90–150°C
3. Low temperature	<90°C
Hot igneous resources (rock intruded in molten form from depth)	
a. Molten material—magma systems	>659°C
b. No molten material—hot dry rock systems	90–650°C
Conduction-dominated resources (heat carried upward by conduction through rock)	
a. Radiogenic (heat generated by radioactive decay)	30–150°C
b. Sedimentary basins (hot fluid in sedimenary rock)	30–150°C
c. Geopressured (hot fluid under high pressure)	150–200°C

   2. *Hot igneous resources*, which include hot dry rock and magma systems

   3. *Conduction-dominated resources*, which include geopressured and radiogenic resources

These basic resource types are distinguished by basic geologic characteristics and the manner in which heat is transferred to Earth's surface, as noted in Table 2. At present only hydrothermal resources are exploited commercially, but research and development activities around the world are developing the potential of the other categories, especially hot dry rock. The following discussion includes a description of and focuses on the general characterization of the features and location of each of these resource categories in the United States.

2.1 The United States Geothermal Resource Base

The U.S. Geological Survey (USGS) compiled an assessment of geothermal resources in the United States in 1975[2] and updated it in 1978[3] which characterizes a *geothermal resource base* for the United States based on geological estimates of all stored heat in the earth above 15°C within six miles of the surface. The defined base ignores the practical *recoverability* of the resource but provides to first order a sense of the scale, scope, and location of the geothermal resource base in the United States.

   The U.S. geothermal resource base includes a set of 108 known geothermal resource areas (KGRAs) encompassing over three million acres in the 11 western states. The USGS resource base captured in these defined KGRAs does not include the lower-grade resource base applicable in direct uses (space heating, greenhouses, etc.), which essentially blankets the entire geography of the nation, although once again ignoring the issues of practical recoverability. Since the 1970s, many of these USGS-defined KGRAs have been explored extensively and some developed commercially for electric power production. For more details see Blair et al.[4]

2.2   Hydrothermal Resources

Hydrothermal convection systems are formed when underground reservoirs carry the Earth's heat toward the surface by convective circulation of steam in the case of *vapor-dominated resources* or water in the case of *liquid-dominated resources*. Vapor-dominated resources are extremely rare on Earth. Three are located in the United States: The Geysers and Mount Lassen in California and the Mud Volcano system in Yellowstone National Park.* All remaining KGRAs in the United States are liquid-dominated resources (located in Figure 1).

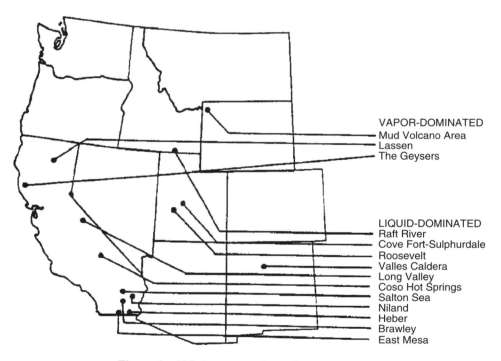

Figure 1   U.S. known geothermal resource areas.

*Other known vapor-dominated resources are located at Larderello and Monte Amiata, Italy, and at Matsukawa, Japan.

Vapor-Dominated Resources

In vapor-dominated hydrothermal systems, boiling of deep subsurface water produces water vapor, which is also often superheated by hot surrounding rock. Many geologists speculate that as the vapor moves toward the surface, a level of cooler near-surface rock induces condensation, which along with the cooler groundwater from the margins of the reservoir, serves to recharge the reservoir. Since fluid convection takes place constantly, the temperature in the vapor-filled area of the reservoir is relatively uniform and a well drilled into this region will yield high-quality* superheated steam, which can be circulated directly in a steam turbine generator to produce electricity.

The most commercially developed geothermal resource in the world today in known as The Geysers in northern California, which is a very high-quality, vapor-dominated hydrothermal convection system. At The Geysers steam is delivered from the reservoir from a depth of 5,000 to 10,000 feet and piped directly to turbine generators to produce electricity. Power production at The Geysers began in 1960, growing to a peak generating capacity in 1987 of over 2000 mWe. Since then it has declined to around 1200 mWe, but still accounts for over 90 percent of the total U.S. geothermal electric generating capacity.

Commercially produced vapor-dominated systems at The Geysers, Lardarello (Italy), and Matsukawa (Japan) all are characterized by reservoir temperatures in excess of 230°C.[†] Accompanying the water vapor in these resources are very small concentrations (less than 5%) of noncondensable gases (mostly carbon dioxide, hydrogen sulfide, and ammonia). The Mount Amiata Field in Italy is actually a different type of vapor-dominated resource, characterized by somewhat lower temperatures than The Geysers-type resource and by much higher concentrations of noncondensable gases. The geology of Mount Amiata-type resources is less well understood than The Geysers-type vapor-dominated resources, but may turn out to be more common because its existence is more difficult to detect.

Liquid-Dominated Resources

Hot-water or wet steam hydrothermal resources are much more commonly found around the globe than dry steam deposits. Hot-water systems are often associated with a hot spring that discharges at the surface. When wet steam deposits occur at considerable depths (also relatively common), the resource temperature is often

---

*High-quality steam is often referred to as dry steam since it contains no entrained liquid water spray. Most steam boilers are designed to produce high-quality steam. Steam with entrained liquid has significantly lower heat content than dry steam. Superheated steam is steam generated at a higher temperature than its equivalent pressure, created either by further heating of the steam (known as superheating), usually in a separate device or section of a boiler known as a superheater, or by dropping the pressure of the steam abruptly, which allows the steam drop to a lower pressure before the extra heat can dissipate.

[†]The temperature of dry steam is 150°C, but steam plants are most cost-effective when the resource temperature is above about 175°C.

well above the normal boiling point of water at atmospheric pressures. These temperatures are know to range from $100°$ to $700°C$ at pressures of 50 to 150 psig. When the water from such resources emerges at the surface, either through wells or through natural geologic anomalies (e.g., geysers), it flashes to wet steam. As noted later, converting such resources to useful energy forms requires more complex technology than that used to obtain energy from vapor-dominated resources.

One of the reasons dealing with wet steam resources is more complex is that the types of impurities found in them vary considerably. Commonly found dissolved salts and minerals include sodium, potassium, lithium, chlorides, sulfates, borates, bicarbonates, and silica. Salinity concentrations can vary from thousands to hundreds of thousands of parts per million. The Wairekei Fields in New Zealand and the Cerro Prieto Fields in Mexico are examples of currently well-developed liquid-dominated resources and in the United States many such resources are in development or under consideration for development.

2.3   Hot Dry Rock and Magma Resources

In some areas of the western United States, geologic anomalies such as tectonic plate movement and volcanic activity have created pockets of impermeable rock covering a magma chamber within six or so miles of the surface. The temperature in these pockets increases with depth and the proximity to the magma chamber, but, because of the impermeability of the rock, they lack a water aquifer. Hence, they are often referred to as *hot dry rock* (HDR) deposits.

A number of schemes for useful energy production from HDR resources have been proposed, but all of them involve creation of an artificial aquifer that is used to bring the heat to the surface. The basic idea is to introduce artificial fractures that connect a production and injection well. Cold water is injected from the surface into the artificial reservoir where the water is heated then returned to the surface through a production well for use in directuse or geothermal power applications. The concept is being tested by the U.S. Department of Energy at Fenton Hill near Los Alamos, New Mexico.

A typical HDR resource extraction system design is shown in Figure 2. The critical parameters affecting the ultimate commercial feasibility of HDR resources are the geothermal gradient throughout the artificial reservoir and the achievable well flow rate from the production well.

Perhaps even more challenging than HDR resource extraction is the notion of extracting thermal energy directly from shallow (several kilometers in depth) magma intrusions beneath volcanic regions. Little has been done to date to develop this kind of resource.

2.4   Geopressured Resources

Near the Gulf Coast of the United States are a number of deep sedimentary basins that are geologically very young, less than 60 million years. In such regions, fluid

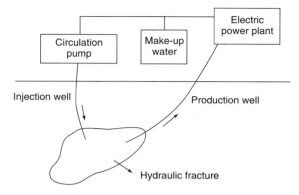

Figure 2 Hot dry rock geothermal resource conversion.

located in subsurface rock formations carry a part of the overburden load, thereby increasing the pressure within the formation. If the water in such a formation is also confined in an insulating clay bed, the normal heat flow of Earth can raise the temperature of the water considerably. The water in such formations is typically of somewhat lower salinity as well, compared to adjacent aquifers, and, in many cases, is saturated with large amounts of recoverable methane. Such formations are referred to as *geopressured* and are considered by some geologists to be promising sources of energy in the coming decades.

The promise of geopressured geothermal resources lies in the fact that they may be able to deliver energy in three forms: (1) mechanical energy, since the gas and liquids are resident in the formations under high hydraulic pressure, (2) the geothermal energy stored in the liquids, and (3) chemical energy, since, as list above, many geopressured resources are accompanied by high concentrations of methane or natural gas.

Geopressured basins exist in several areas within the United States, but those considered the most promising are located in the Texas–Louisiana coast. They are of particular interest because they are very large in terms of both areal extent and thickness and because the geopressured liquids (mostly high-salinity brine) are suspected to include high concentrations of methane.

In past evaluations of the Gulf Coast region, a number of *geopressured fairways* were identified, which are thick sandstone bodies expected to contain geopressured fluids of at least 150°C. Detailed studies of the fairways of the Frio Formation in East Texas were completed in 1979, although only one, Brazoria, met the requirements for further well testing and remains the subject of interest by researchers.

3 GEOTHERMAL ENERGY CONVERSION

For modern society, geothermal energy has a number of important advantages. Although not immediately renewable like solar and wind resources, Earth's energy is vast and essentially inexhaustible (i.e., with a lifetime of billions of years); environmental impacts associated with geothermal energy conversion are generally modest and local compared with other alternatives; and energy production is generally very reliable and available day and night. In addition, geothermal energy is not generally affected by weather, although there may be seasonal differences in plant efficiency. Finally, geothermal plants take little space and can be made unobtrusive even in areas of high scenic value, where many geothermal resources are located.

Most economical applications of geothermal energy, at least at this point in the development of the necessary technology, hinge on the availability and quality of the resource. On one hand, geothermal resources are far less pervasive than solar or wind resources. On the other hand, as technology continues to develop, the use of lower-quality but much more common geothermal resources may increase their development substantially.

Since use of geothermal energy involves interaction with a geologic system, the characteristics and quality of the resource involves some natural variability (far less than with solar or wind), but, more importantly, the utilization of the geothermal resource can be affected profoundly by the way in which the resource is tapped. In particular, drawing steam or hot water from a geothermal aquifer at a rate higher than the rate at which the aquifer is refreshed will reduce the temperature and pressure of the resource available for use locally and can precipitate geologic subsidence evident even at the surface. The consequences of resource utilization for the quality of the resource are especially important since geothermal energy is used immediately and not stored, in contrast to the case of oil and gas resources, and would undermine the availability, stability, and reliability of the commercially produced energy from the resource and diminish its value.

Reinjecting geothermal fluids that remain after the water (or steam) has been utilized in a turbine (or other technology that extracts the useful heat from the fluid) helps preserve the fluid volume of the reservoir and is now a common practice for environmental reasons and to mitigate subsidence.* Nonetheless, even with reinjection, the heat content of a well-developed geothermal reservoir will gradually decline, as typified by the history at The Geysers.

A variety of technologies are in current use to convert geothermal energy to useful forms. These can very generally be grouped into three basic categories: (1) direct use, (2) electric power generation, and (3) geothermal heat pumps. Each category utilizes the geothermal resource in a very different way.

---

*Reinjection of water is common in oil and gas field maintenance to preserve the volume and pressure of the resource in those fields and the basic concept is applicable in geothermal fields as well to both mitigate the environmental impacts of otherwise disposing of spent geothermal liquids and to maintain the volume, temperature, and pressure of the geothermal aquifer.

3.1 Direct Uses of Geothermal Energy

The heat from geothermal resources is frequently used directly without a heat pump or to produce electric power. Such applications generally use lower-temperature geothermal resources for space heating (commercial buildings, homes, greenhouses, etc.), industrial processes requiring low-grade heat (drying, curing, food processing, etc.), or aquaculture.

Generally, these applications use heat exchangers to extract the heat from geothermal fluids delivered from geothermal wells. Then, as noted earlier, the spent fluids are then injected back into the aquifer through reinjection wells. The heat exchangers transfer the heat from the geothermal fluid usually to fresh water that is circulated in pipes and heating equipment for the direct use applications.

Such applications can be very efficient in small end-use applications such as greenhouses, but it is generally necessary for the applications to be located close to the geothermal heat source. Perhaps the most spectacular and famous example of direct use of geothermal energy is the city of Reykjavik, Iceland, which is heated almost entirely with geothermal energy.

3.2 Electric Power Generation

Geothermal electric power generation generally uses higher-temperature geothermal resources (above $110°C$). The appropriate technology used in power conversion depends on the nature of the resources.

As noted earlier, for vapor-dominated resources, it is possible to use direct steam conversion. For higher-quality liquid-dominated hydrothermal resources—with temperatures greater than $180°C$, power plants can be used to separate steam (flashed) from the geothermal fluid and then feed the steam into a turbine that turns a generator. For lower-quality resources so-called *binary* power plants can increase the efficiency of electric power production from liquid-dominated resources.

In a manner similar to direct uses of geothermal energy, binary power plants use a secondary working fluid that is heated by the geothermal fluid in a heat exchanger. In binary power plants, however, the secondary working fluid is usually a substance such as isobutane, which is easily liquified under pressure but immediately vaporizes when the pressure is released at lower temperatures than that of water. Hence, the working-fluid vapor turns the turbine and is condensed prior to reheating in a heat exchanger to form a closed-loop working cycle.

In all versions of geothermal electric power generation, the spent geothermal fluids are ultimately injected back into the reservoir. Geothermal power plants vary in capacity from several hundred kWe to hundreds of mWe. In the United States, at the end of 2002, there were 43 geothermal power plants, mostly located in California and Nevada. In addition, Utah has two operating plants and Hawaii has one. The power-generating capacity at The Geysers remains the largest concentration of geothermal electric power production in the world, producing almost as much electricity as all the other U.S. geothermal sites combined.

Muffler[3] estimates that identified hydrothermal resources in the United States could provide as much as 23,000 mWe of electric generating capacity for 30 years, and undiscovered hydrothermal resources in the nation could provide as much as five times that amount. Kutscher[5] observes that if hot dry rock resources become economically recoverable in the United States, they would be "sufficient to provide our current electric demand for tens of thousands of years," although currently economically tapping hot dry rock resources remains largely elusive and speculative. To explore that potential, a variety of research and development program activities are underway sponsored by the U.S. government.[1]

The following sections explore more specifically the technologies of direct steam, flash, and binary geothermal energy conversion along with the strategy of combining geothermal energy with fossil (oil, coal, or natural gas) in power generation.

Direct Steam Conversion

Electric power generation using the geothermal resources at The Geysers in California and in central Italy, which were referred to earlier as The Geysers-type vapor-dominated resources, is a very straightforward process relative to the processes associated with other kinds of geothermal resources. A simplified flow diagram of direct steam conversion is shown in Figure 3. The key components of such a system include the steam turbine–generator, condenser, cooling towers, and some smaller facilities for degassing and removal of entrained solids and for pollution control of some of the noncondensable gases.

The process begins when the naturally pressurized steam is piped from production wells to a power plant, where it is routed through a turbine generator to produce electricity. The geothermal steam is supplied to the turbine directly,

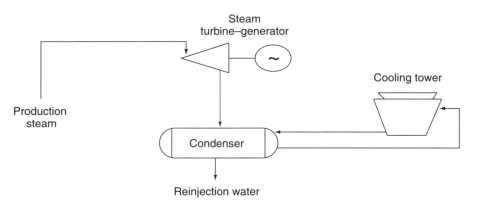

Figure 3   Direct steam conversion.

except for the relatively simple removal of entrained solids in gravity separators or the removal of noncondensable gases in degassing vessels. Such gases include carbon dioxide, hydrogen sulfide, methane, nitrogen, oxygen, hydrogen, and ammonia. In modern geothermal plants additional equipment is added to control, in particular, the hydrogen sulfide and methane emissions from the degassing stage. Release of hydrogen sulfide is generally recognized as the most important environmental issue associated with direct steam conversion plants at The Geysers' generating facilities. The most commonly applied control technology for abatement of toxic gases such as hydrogen sulfide in geothermal power plants is known as the Stratford process.

As the *filtered* steam from the gravity separators and degassing units expands in the turbine it begins to condense. It is then exhausted to a condenser, where it cools and condenses completely to its liquid state and is subsequently pumped from the plant. The condensate is then almost always reinjected into the subterranean aquifer at a location somewhat removed from the production well. Cooling in the condenser is provided by a piping loop between the condenser and the cooling towers. The hot water carrying the heat extracted from the condensing steam line from the turbine is routed to the cooling tower where the heat is rejected to the atmosphere. The coolant fluid, freshly cooled in the cooling tower, is then routed back to the condenser, forming the complete cooling loop (as shown in Figure 3).

Reinjection of geothermal fluids in modern geothermal systems is almost always employed to help preserve reservoir volume and to help mitigate air and water pollutant emissions on the surface. However, as the geothermal well field is developed and the resource produced, effective reservoir maintenance becomes an increasingly important issue. For example, in The Geysers, noted earlier as a highly developed geothermal resource, as the geothermal fluids are withdrawn and reinjected, the removal of the heat used in power generation causes the reservoir temperature to decline. The cooling reservoir then contracts, and this is observed at the surface as subsidence.* Geophysicists Mossop and Segall observe that subsidence at The Geysers has been on the order of 0.05 m per year since the early 1970s.[6]

Because of the quality of the resource and the simplicity of the necessary equipment, direct steam conversion is the most efficient type of geothermal electric power generation. A typical measure of plant efficiency is the amount of electric energy produced per pound of steam at a standard temperature (usually around 175°C). For example, the power plants at The Geysers produce 50–55 Whr of electricity per pound of 176°C steam used, which is a very high-quality geothermal resource. Another common measure of efficiency is known as the geothermal

---

*Most researchers conclude that the extraction, reinjection, and associated temperature decline causes strain due to a combination of thermoelastic and poroelastic deformations, which results in surface subsidence. See, for example, Mossop and Segall in Ref. 6.

resource utilization efficiency (GRUE), defined as the ratio of the net power output of a plant to the difference in the thermodynamic availability of the geothermal fluid entering the plant and that of that fluid at ambient conditions. Power plants at The Geysers operate at a GRUE of 50 to 56 percent.

Flashed Steam Conversion

Most geothermal resources do not produce dry steam, but rather a pressurized two-phase mixture of steam and water often referred to as *wet steam*. When the temperature of the geothermal fluid in this kind of resource regime is greater than about 180°C, plants can use the flashed steam energy conversion process. Figure 4 is a simplified schematic that illustrates the flashed steam power generation process used in such plants. In addition to the key components used in direct steam conversion plants (i.e., turbine, condenser, and cooling towers), flashed steam plants include a component called a separator or flash vessel.

The flash conversion process begins with the geothermal fluid flows from the production well(s) flows under its own pressure into the separator, where saturated steam is flashed from the liquid brine. That is, as the pressure of the fluid emerging from the resource decreases in the separator, the water boils or *flashes* to steam and the water and steam are separated. The steam is diverted into the power production facility and the spent steam and remaining water are then reinjected into the aquifer.

Many geothermal power plants use multiple stages of flash vessels to improve the plant efficiency and raise power generation output. Figure 5 is a simplified

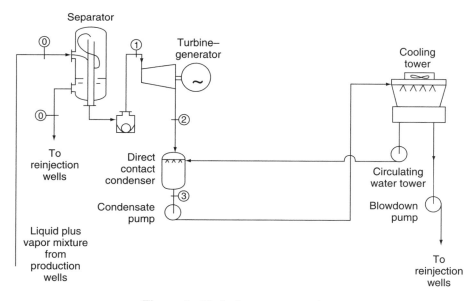

Figure 4 Flashed steam conversion.

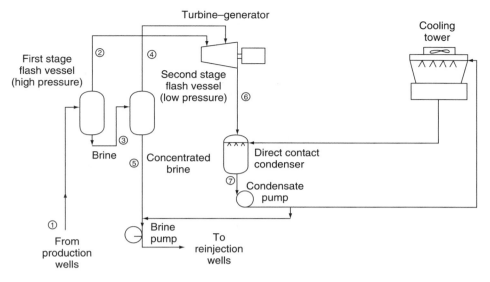

Figure 5   Two-stage flash conversion.

schematic illustrating a two-stage or *dual-flash system*. Such systems are designed to extract additional energy from geothermal resource by capturing energy from both high and lower temperature steam.

In the two-stage process, the unflashed fluid leaving the initial flash vessel enters a second flash vessel that operates at a lower pressure, causing additional steam to be flashed. This lower-pressure steam is supplied to the low-pressure section of the steam turbine, recovering energy that would have been lost if a single-stage flash process had been used. The two-stage process can result in a 37 percent or better improvement in plant performance compared with a single-stage process. Additional stages can be included as well, resulting in successively diminishing levels of additional efficiency improvement. For example, addition of a third stage can add an additional 6 percent in plant performance.

Binary Cycle Conversion

For lower-quality geothermal resource temperatures—usually below about $175°C$,—flash power conversion is not efficient enough to be cost effective. In such situations, it becomes more efficient to employ a binary cycle. In the binary cycle, heat is transferred from the geothermal fluid to a volatile working fluid (usually a hydrocarbon such as iobutane or isopentane) that vaporizes and is passed through a turbine. Such plants are called binary since the secondary fluid is used in a Rankine power production cycle, and the primary geothermal fluid is used to heat the working fluid. These power plants generally have higher equipment costs than flash plants because the system is more complex.

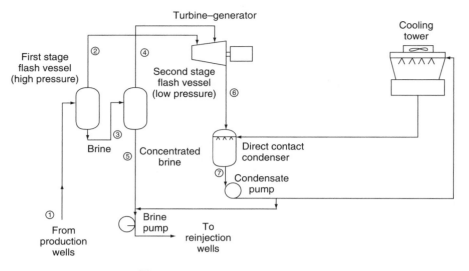

Figure 6   Binary cycle conversion.

Figure 6 is a simplified schematic illustrating the key components in the binary cycle conversion process. Geothermal brine from the production well(s) passes through a heat exchanger, where it transfers heat to the secondary working fluid. The cooled brine is then reinjected into the aquifer. The secondary working fluid is vaporized and superheated in the heat exchanger and expanded through a turbine, which drives an electric generator. The turbine exhaust is condensed in a surface condenser, and the condensate is pressurized and returned to the heat exchanger to complete the cycle. A cooling tower and a circulating water system reject the heat of condensation to the atmosphere.

A number of variations of the binary cycle have been designed for geothermal electric power generation. For example, a regenerator may be added between the turbine and condenser to recover energy from the turbine exhaust for condensate heating and to improve plant efficiency. The surface-type heat exchanger, which passes heat from the brine to the working fluid, may be replaced with a direct contact or fluidized-bed type exchanger to reduce plant cost. Hybrid plants combining the flashed steam and binary processes have also been evaluated in many geothermal power generation applications.

The binary process is proving to be an attractive alternative to the flashed steam process at geothermal resource locations that produce high-salinity brine. First, since the brine can remain in a pressurized liquid state throughout the process and does not pass through the turbine, problems associated with salt precipitation and scaling as well as corrosion and erosion can be greatly reduced. In addition, binary cycles offer the additional advantage that a working fluid can be selected that has superior thermodynamic characteristics to steam, resulting in a more efficient conversion cycle. Finally, because all the geothermal brine is reinjected into the aquifer, binary cycle plants do not require mitigation of gaseous emissions, and reservoir fluid volume is maintained. Larger binary plants are typically constructed as a series of smaller units or modules, so maintenance can be completed on individual modules without shutting down the entire plant, thereby minimizing the impact on total plant output.

Some Additional System Selection Considerations

The overall efficiency of energy conversion processes for liquid-dominated resources is dependent primarily on the resource temperature and to a lesser degree on brine salinity and the concentration of noncondensable gases. System efficiency can generally be improved by system modifications, but such modifications usually involve additional cost and complexity. Figure 7 shows an empirical family of curves relating power production per unit of geothermal brine consumed for both two-stage flash and binary conversion systems.

The level of hydrogen sulfide emissions is an important consideration in geothermal power plant design. Emissions of hydrogen sulfide at liquid-dominated geothermal power plants are generally lower than for direct steam processes. For example, steam plants emit 30 to 50 percent less hydrogen sulfide than direct steam plants. Binary plants would generally not emit significant amounts of hydrogen sulfide because the brine remains contained and pressurized throughout the entire process.

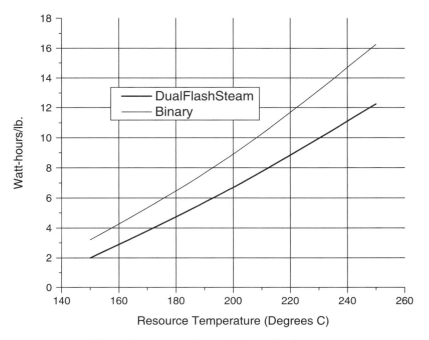

Figure 7   Net geothermal brine effectiveness.

Finally, the possibility of land surface subsidence caused by withdrawal of the brine from the geothermal resource can be an important design consideration. Reinjection of brine is the principal remedy for avoiding subsidence by maintaining reservoir volume and has the added environmental benefit of minimizing other pollution emissions to the atmosphere. However, faulty reinjection can contaminate local fresh groundwater. Also in some plant designs if all brine is reinjected, an external source of water is required for plant cooling water makeup.

Hybrid Geothermal/Fossil Energy Conversion

Hybrid fossil/geothermal power plants use both fossil energy and geothermal heat to produce electric power. A number of alternative designs exist. First, a geothermal preheat system involves using geothermal brine to preheat the feedwater in an otherwise conventional fossil-fired power plant. Another variation is fossil superheat concept that incorporates a fossil-fired heater to superheat geothermal steam prior to expansion in a turbine.

3.3 Geothermal Heat Pumps

Geothermal heat pumps (GHP), sometimes also referred to as groundwater heat pumps, use the earth's typical diffuse low-grade heat found in the very shallow subsurface (usually between 30 and 300 ft in depth), usually in space-heating applications. In most geographic areas in the United States, GHP can deliver three to four times more energy than it consumes in the electricity needed to operate and can be used over a wide range of earth temperatures.

The GHP energy-conversion process works much like a refrigerator, except that it is reversible, that is, the GHP can move heat either into the earth for cooling or out of the earth for heating, depending on whether it is summer or winter. GHP can be used instead of or in addition to direct uses of geothermal energy for space or industrial process heating (or cooling), but the shallow resource used by GHP is available essentially anywhere, constrained principally by land use and economics, especially initial installation costs.

The key components of the GHP system include a ground refrigerant-to-water heat exchanger, refrigerant piping and control valves, a compressor, an air coil (used to heat in winter and to cool and dehumidify in summer), a fan, and control equipment. This system is illustrated in Figure 8.

The GHP energy-conversion process begins with the ground heat exchanger, which is usually a system of pipes configured either as a closed-or open-loop system. The most common configuration is the closed loop, in which high-density polyethylene pipe is buried horizontally at a depth of at least 4 to 6 ft deep or vertically at a depth of 100 to 400 ft. The pipes are typically filled with a refrigerant solution of antifreeze and water, which acts as a *heat exchanger*. That is, in winter, the fluid in the pipes extracts heat from the earth and carries it

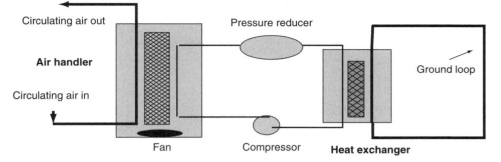

Circulating air out   Pressure reducer

Air handler   Ground loop

Circulating air in

Fan   Compressor   **Heat exchanger**

Figure 8   Geothermal heat pump system configuration.

into the building. In the summer, the system reverses the process and takes heat from the building and transfers it to the cooler ground.

GHP systems deliver heated or cooled air to residential or commercial space through ductwork just like conventional heating, ventilating, and air conditioning (HVAC) systems. An indoor coil and fan called an *air handler* also contains a large blower and a filter just like conventional air conditioners.

Ground-Loop GHP Systems

There are four basic types of *ground-loop GHP systems*: (1) horizontal, (2) vertical, (3) pond/lake, and (4) open-loop configurations; the first three of these are all closed-loop systems. Selection of one of these system types depends on climate, soil conditions, available land, and local installation costs. The following briefly describes each of the approaches to GHP ground-loop systems.

- *Horizontal.* Considered generally most cost-effective for residential installations, especially for new construction where sufficient land is available, the installation entails two pipes buried in trenches that form a loop.
- *Vertical.* Vertically-oriented systems are often used for large commercial buildings and schools where the land area required for horizontal loops would be prohibitive or where the surrounding soil is too shallow for trenching in a horizontal system or when a goal is to minimize the disturbance to existing landscaping. In such systems holes are drilled about 20 ft apart and 100 to 400 ft deep and pipes are installed and connected at the bottom to form the loop.
- *Pond/lake.* If the site has a suitable water body accessible, a supply line pipe can be run from the building to the water and coiled under the surface of the water body to prevent freezing in winter.
- *Open-loop.* An open-loop system uses water from well(s) or a surface body of water as the heat exchange fluid that circulates directly through the GHP system. Once the water has circulated through the heat exchanger, the water

returns to the ground through another well or by surface discharge. This option can be used only where there is an adequate supply of relatively clean water and where its use is permitted under local environmental codes and regulations.

REFERENCES

1. U.S. Department of Energy, Office of Energy Efficiency and Renewable Energy, Strategic Plan, DOE/GO-102002-1649, (October 2002).

2. White, D. E., and D. L. Williams (eds.), "Assessment of Geothermal Resources of the United States—1975," U.S. Geological Survey, Arlington, VA, Circular 726, 1975.

3. Muffler, L. J. P. (ed.), "Assessment of Geological Resources of the United States—1978," U.S. Geological Survey, Arlington, VA, Circular 790, 1979.

4. Blair, P. D., T. A. V. Cassel, and R. H. Edelstein, *Geothermal Energy: Investment Decisions and Commercial Development*, Wiley, New York, 1982.

5. Kutscher, C. F., "The Status and Future of Geothermal Electric Power," NREL/CP-550-28204, National Renewable Energy Laboratory, Golden, CO, August 2000.

6. Mossop, A., and P. Segall, "Subsidence at the Geysers Geothermal Field," *Geophysical Research Letters*, 24(14), 1839–1842 (1997).

CHAPTER 5

WIND POWER GENERATION

Todd S. Nemec
GE Energy
Schenectady, New York

1	MARKET AND ECONOMICS	119	4	ROTOR AND DRIVE TRAIN DESIGN	124
2	CONFIGURATIONS	120	5	SITE SELECTION	126
3	POWER PRODUCTION AND ENERGY YIELD	121			

1 MARKET AND ECONOMICS

Wind power has long been used for grain-milling and water-pumping applications. Significant technical progress since the 1980s, however, driven by advances in aerodynamics, materials, design, controls, and computing power, has led to economically competitive electrical energy production from wind turbines. Technology development, favorable economic incentives (due to its early development status and environmental benefits), and increasing costs of power from traditional fossil sources have led to significant worldwide sales growth since the early 1980s. Production has progressed at an even faster pace beginning in the late 1990s. Figure 1, below shows the U.S. wind turbine installations (MW, net) since 1981.

The spike in U.S. wind turbine installations from 1982 to 1985 was due to generous tax incentives (up to 50 percent in California[2]), access to excellent wind resources, and high fossil-fuel prices. Today, Germany, the United States, Spain, and Denmark lead in installed MW, although significant growth is occurring worldwide.[3] From an energy-share standpoint, the northern German state, Shleswig-Holstein, produces approximately 30 percent of its electric energy from wind power, while Denmark produces about 20 percent.[4]

Like other power-producing technologies, wind turbines are measured on their ability to provide low cost of electricity (COE) and high project net present value (NPV). Unlike fossil plants, however, fuel (wind energy) is free. This causes COE to be dominated by the ratio of costs per unit energy, rather than a combination of capital costs, fuel cost, and thermal efficiency. For customers purchasing based

119

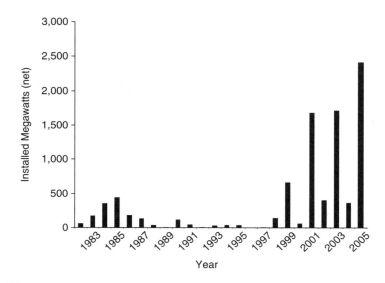

Figure 1   U.S. wind turbine installations (MW) by year. (From Ref. 1.)

on highest NPV, high power sale prices and energy production credits can drive turbine optimization to a larger size (and/or energy capture per rated MW), and higher COE, than would be expected from a typical optimization for lowest COE. Operation and maintenance costs (in cents per kWh) for wind turbines trend higher than those for fossil plants, primarily due to their lower power density. The ability to predict and trade life-cycle costs versus energy improvement, from new technologies, is a key contributor to efficient technology development and market success.

2   CONFIGURATIONS

The most popular configuration for power-generating wind turbines is the upwind three-bladed Horizontal Axis Wind Turbine (HAWT), shown in Figure 2. *Upwind* refers to the position of the blades relative to the tower.

Wind turbine configurations can be traced back to vertical-axis drag-type machines used for milling grain, which had the theoretical potential to achieve an 8 percent power coefficient, or percent energy extracted from the wind.[6] Modern Vertical Axis Wind Turbines (VAWT), like HAWTs, use the much more effective *lift principle* to produce power. VAWTs have been built in both Darrieus (curved blades connected at one or both ends) and H (separate vertical blades; also called giromill) configurations, although neither has been put into widespread use. VAWT aerodynamics are somewhat more complex, with a constantly changing angle of attack, and analyses have generally concluded that their power coefficient entitlement is lower than HAWTs.[7] Figure 3 shows the nacelle cutaway view of a horizontal-axis turbine.

Figure 2 GE 2 · X prototype in Wieringerneer, Netherlands. (Courtesy of Ref. 5.)

The rotor, made up of the blades and hub, rotates a drive train through the low-speed shaft connected to a gearbox, high-speed shaft, and generator (or from the low-speed shaft to a direct-drive generator). The nacelle consists of the base frame and enclosure; it houses the drive train, various systems, and electronics required for turbine operation. Towers are made of steel or steel-reinforced concrete. Steel towers use either a tubular or lattice type construction. Today's turbine configuration has evolved from both scaling-up and adding features to small wind turbine designs, and from private and government-sponsored development of large machines.

3 POWER PRODUCTION AND ENERGY YIELD

Turbines extract energy from the wind according to following formula, derived from the first law of thermodynamics:

$$P = C_p \tfrac{1}{2} \rho A U^3$$

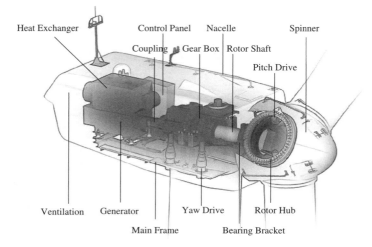

Heat Exchanger Control Panel Nacelle Spinner
Coupling Gear Box Rotor Shaft
Pitch Drive
Ventilation Generator Yaw Drive Rotor Hub
Main Frame Bearing Bracket

Figure 3 GE 1.5s nacelle, cutaway view. (Courtesy of Ref. 5.)

where

P = Power
C_p = Power coefficient
ρ = Air density
A = Rotor swept area
U = Air velocity at hub height

This equation shows power to be a function of air density and swept area, while varying by the cube of wind speed. These functions are not exact in real calculations, however, as aerodynamic and drive-train characteristics restrict power coefficient over much of the operating range. The maximum theoretical power coefficient with zero airfoil drag and other simplifying assumptions is 59.3 percent, while modern turbines deliver peak coefficients in the mid-40 percent range.

The peak efficiency corresponds to a rotor exit air velocity of one-third the initial wind speed. This wake effect—along with site geographic, turbulence, and wind rose data—is significant when planning turbine spacing and arrangement on a multi-turbine wind farm. Turbulence acts to reduce the velocity reduction immediately behind the turbine by re-energizing the wake, while it also spreads the energy loss over larger area.[6] Crosswind spacing, depending on wind characteristics, can usually be much closer than downwind spacing—crosswind tower spacing is on the order of three to five rotor diameters.

The power equation also provides insight into the basic power and mass scaling relationships. Power increases as a function of area, a function of diameter squared, while mass is a function of volume, or diameter cubed. This is true for aerodynamically load-limited components. Most electrical capacities and costs

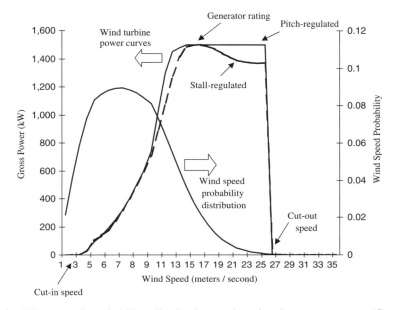

Figure 4 Wind speed probability distribution and notional power curves. (Courtesy of Ref. 5.)

scale with rated power, while some part sizes are independent of turbine size. The cubed-squared relationship between component mass and power is the same found in most power-generation cycles, such as gas turbines. Larger wind turbines are made economically possible by both reducing this 3/2 exponent, by improving technology and design strategy, such as using more advanced materials, and from increased leverage over fixed costs.

Wind turbine performance is characterized by its power curve, Figure 4, which shows the gross power produced as a function of wind speed. This curve assumes clean airfoils, standard control schedules, a given wind turbulence intensity, and a sea-level air density. Three items to note on the power curve are the cut-in speed, cut-out speed, and generator rating. Cut-in speed is determined by the wind speed where the aerodynamic torque is enough to overcome losses. Cut-out speed is set to balance the power production in high winds with design loads and costs. Both speeds have dead-band regions around them to minimize the number of start–stop transients during small changes in wind speed.

Like cut-out speed, selecting the rotor diameter relative to generator rating requires balancing higher energy production at high wind speeds while minimizing costs.

Economic and design analysis has proven that turbines designed for high wind-speed operation should have a larger generator relative to rotor size, while those designed for low wind speed will have a larger rotor for a given generator size.[8] Advances in design and controls technology have not only helped turbines scale economically to larger sizes, but have allowed turbines to run larger rotor diameters at a given rating.[4]

Representative power curves are shown for two methods of limiting power in high winds. The pitch-controlled blade curve shows a constant generator output above rated power, while the curve for a stall-controlled (fixed-pitch) turbine delivers a peaked profile. Gross annual energy production (in kilowatt-hours, kWh) is calculated by multiplying the wind probability (in annual hours) at each wind speed by the power curve KW at that same wind speed, then adding up the total. Because of the higher probability for low wind speeds and turbine design/economic tradeoffs, wind turbines operate at 25 percent to 45 percent plus net capacity factor, depending on the turbine and site. Capacity factor is the fraction of energy produced relative to the rated capacity. This is much lower than the 95 percent plus levels achieved by dispatchable fossil-fuel power plants.

Annual energy yield = 5,000,000 kWh (assumed: measured from a kWh meter,

or calculated based on turbine and wind conditions)

Rated capacity = 1,500 kW (Rating) × 8,760 hours/year

Capacity Factor = 5,000,000/(1,500 × 8,760) = 38%

Some of the gross-to-net loss is bookkept as availability losses (1 to 4 percent of energy produced), which are caused by both forced and scheduled outages. Other losses can total less than 15 percent and include array interference effects, electrical collection losses, blade soiling, and control losses.

Predictive performance analysis generally assumes that wind speed probability follows a Rayleigh distribution (Weibull distribution with shape factor equal to 2), along with an average wind speed at the hub height. In-depth and site analyses will use modifications to this statistical model, or will use data unique to a given site. Although the previous example considers the effects of vertical wind shear on average wind speed at hub-height, detailed energy yield analysis and loads calculations need to consider the effects of vertical velocity distribution. Taller towers allow turbines to see a higher average wind speed due to reduced friction with the ground and other objects at lower heights.

4 ROTOR AND DRIVE TRAIN DESIGN

Rotor and drive train are ultimately optimized to yield the best economics for the turbine's mission. This is part of a multidisciplinary process involving aerodynamics, weight, materials, aeroelasticity, life, first cost, operating cost, frequency response, controls strategy, configuration options/technology availability, noise, site characteristics, supply chain, and a customer value equation. One modern mission requirement is quieter operation for land-based turbines—various noise sources correlate with tip speed raised to powers as high as the five, among other variables.[6] Aerodynamic characteristics that are selected include number of blades, tip speed ratio, blade radius, solidity, blade twist, chord length, airfoil section, and so on in greater detail. As an example of this process, consider that at a given rated power, higher tip speed ratio (higher rotor speed at a given wind speed) does the following:

- It reduces main shaft torque requirements and component sizes (costs).
- It increases noise.
- It decreases rotor solidity (to maintain or increase the power coefficient), which reduces blade chord length and thickness (for a given number of blades).
- It makes it more difficult to fabricate blades and achieve strength objectives.

Prior to generating power, wind turbines were generally configured as direct-drive units to pump water or mill grain. These applications place a high emphasis on torque coefficient at zero rotor speed, defining their ability to start under load. A high-power coefficient was sacrificed by using high blade solidity (blade area divided by rotor disk area) and low tip-speed ratios (tip speed divided by wind speed) in order to achieve high torque.[9] When transferring power through an electrical connection, such as a generator, rotor design generally favors higher power coefficients via low solidity and high tip-speed ratio, resulting in modern high aspect-ratio blade shapes. Three blades have generally been favored over two because power coefficient is higher at lower tip-speed ratios, and for several structural dynamic considerations: out-of-plane bending loads are higher on two-blade designs due to wind shear, tower shadow effect, effects of an upward-tiled shaft (to improve tower clearance), and from yaw-induced moments due to a changing moment of inertia.[10] Several of these loads can be eliminated in two-blade rotors with the use of a teetering hub.

As already described above, a wind turbine will generally be designed with little excess weight, or structural design, margin in order optimize life-cycle economics. Design misses, such as higher weight blades, have subsequent effect in weight, cost, and/or life of drive train, tower, and foundation components. Turbine fatigue and ultimate loads are driven by four categories: aerodynamic, gravity, dynamic interactions, and control.[11]

Drive trains absorb the rotor loads and distribute them to the bedplate for transmission to the tower and foundation. They also serve to convert torque into electrical power via the nacelle-mounted generator. Direct-drive generators, used by some manufacturers, turn at the rotor RPM, use a higher number of generator poles, and use power electronics to convert this rotor RPM into 50 Hz or 60 Hz AC current. Geared drive trains use a gearbox to drive a high-speed shaft connected to a smaller generator (with a fewer number of poles). Most manufacturers are employing variable-speed and pitch control using power electronics, pitch controllers, gearboxes, and induction (asynchronous) generators to optimize cost, energy yield, and improve grid power quality.

Design advances are evident in the lighter and more compact drive trains. ENERCON GmbH has used direct-drive generators since the early 1990s.[12] These, and mixed solutions that use a single-stage gearbox to step up to a smaller low-speed generator, have been receiving more attention by other manufacturers as power electronic costs have come down.

5 SITE SELECTION

Turbine siting tasks are designed to solve a wide range of economic, environmental, social, and technical issues. Computer modeling of wind-farm concepts can help estimate both the wind resource as well as improve understanding of visual, acoustic, and environmental issues. Some of the early site election activities include the following:

1. Wind resource
   - Determining location(s) with highest average wind speeds
   - Estimating array losses and terrain effects
2. Revenue
   - Energy
   - Capital, energy, and/or emissions incentives
3. Costs
   - Transportation and construction access
   - Grid interconnection costs and transmission impact
   - Land-lease and/or opportunity costs
   - Foundation costs and geological compatibility
4. Site access and environmental
   - Noise and visual restrictions
   - Access rights
   - Impact on wildlife such as birds, bats, or endangered species
   - Interference with aviation flight routes or radar

Micrositing optimizes turbine placement at a given site through the detailed evaluation of energy resource and iteration for best energy yield and/or farm economics. It can be performed after or during the early selection process. State-of-the-art micrositing utilizes macro- and micro-level weather and flow models that are correlated to both long-term (usually low-resolution data, such as airport weather stations) and shorter-term high-resolution data taken from meteorological masts. Models will include topographical features and turbulence estimates, and should be able to produce uncertainty estimates that are useful in financial risk calculations. Ideally, micrositing optimization will include the impact of cost, such as roads and electrical collection, and noise/control strategy in addition to energy yield.

Although it pays for large farms to expend considerable resources to optimize farm layout, smaller installations consisting of one or a few small turbines may not want to cover the cost of detailed analysis. For these cases, rules of thumb can be used to optimize siting to account for turbulence and boundary-layer effects caused by surface roughness and interference, topographical features in the terrain, and turbine wakes. Boundary layer impact on wind velocity is usually expressed by a power-law equation, often using a default one-seventh exponent to model a typical vertical wind shear profile:

$$V(z)/V(z_r) = (z/z_r)^\alpha$$

where

$\alpha = 1/7 = 0.143$
$V(z) =$ Average wind velocity (m/s) at hub height
$V(z_r) =$ average wind velocity (m/s) at reference elevation
$z =$ elevation, m
$z_r =$ reference elevation, m

This allows correcting from a measured (reference) wind location, such as a 10 m weather tower to a much taller wind turbine hub-height. The actual exponent will be calculated from meteorological mast data, extending to a much taller height, and will vary with wind direction and speed (topography, array interference), among other factors. For improved—but still approximate—calculations, the exponent equation can be replaced with an expression based on terrain features:

$$V(z)/V(z_r) = \ln(z/z_0)/\ln(z_r/z_0)$$

where

$V(z) =$ Average wind velocity (m/s) at hub height
$V(z_r) =$ average wind velocity (m/s) at reference elevation
$z =$ elevation, m
$z_0 =$ roughness length, m
$z_r =$ reference elevation, m

Terrain	z_0, Roughness Length (m)
Cities, forests	0.7
Suburbs, wooded countryside	0.3
Villages, countryside with trees & hedges	0.1
Open farmland	0.03
Flat, grassy plains	0.01
Flat desert, rough sea	0.001
Calm open sea	0.0002

As both equations show, wind speed will be higher at a given hub height when there is reduced interference or vertical shear. This is favorable for energy yield, allowing lower hub heights to collect the same wind energy, reducing tower, foundation, and installation costs. Expected revenue should be weighed versus these costs to guide the micrositing turbine placement and farm design.

REFERENCES

1. Wind Energy Program and AWEA Global Market Reports: U.S. DOE, Office of Energy Efficiency and Renewable Energy, Wind and Hydropower Technologies Program and AWEA data, Retrieved October 11, 2006 from the U.S. DOE website: http://www.eere.energy.gov/windandhydro/windpoweringamerica/wind_wind_installed_capacity.asp

2. Mark Gielecki, et al., "Renewable Energy 2000, Issues and Trends: Incentives, Mandates, and Government Programs for Promoting Renewable Energy," U.S. DOE Energy Information Association (EIA), 2001.

3. Joint EWEA/AWEA press release, "Global Wind Power Growth Continues to Strengthen," March 10, 2004.

4. European Wind Energy Association (EWEA), "Wind Energy—The Facts; An Analysis of Wind Energy in the EU-25," 2003, Retrieved from EWEA Web site: www.ewea.org on October 15, 2004.

5. General Electric, Schenectady, NY, 2004.

6. David Spera, ed., *Wind Turbine Technology*, ASME Press, New York, 1995.

7. Robert Harrison, et al., *Large Wind Turbines, Design and Economics*, John Wiley, New York 2000.

8. D. J. Malcolm and A. C. Hansen, "WindPACT Turbine Rotor Design, Specific Rating Study," National Renewable Energy Laboratory (NREL) 2003.

9. J. A. C. Kentfield, *The Fundamentals of Wind-Driven Water Pumps*, Gordon and Breach Science Publishers, Amsterdam, 1996.

10. Tony Burton, et al., *Wind Energy Handbook*, John Wiley, New York, 2001.

11. J. F. Manwell, et al., *Wind Energy Explained, Theory, Design and Application*, John Wiley, West Sussex, England, 2003.

12. ENERCON History, Retrieved October 21, 2004, from Enercon Web site: www.enercon.de.

CHAPTER 6

COGENERATION

Jerald A. Caton
Department of Mechanical Engineering
Texas A&M University
College Station, Texas

1	**INTRODUCTION**	**130**
	1.1 History of Cogeneration	130
	1.2 Constraints on Cogeneration	131
2	**BASIC COGENERATION SYSTEMS**	**132**
	2.1 Topping Cycles	132
	2.2 Bottoming Cycles	133
	2.3 Combined Cycles	134
	2.4 Applications of Cogeneration Systems	135
3	**DESCRIPTIONS OF PRIME MOVERS**	**135**
	3.1 Steam Turbines	136
	3.2 Gas Turbines	136
	3.3 Reciprocating Engines	138
	3.4 Other Possible Prime Movers	139
4	**DESCRIPTION OF OTHER EQUIPMENT AND COMPONENTS**	**139**
	4.1 Electrical Equipment	139
	4.2 Heat-Recovery Equipment	140
	4.3 Absorption Chillers	142
	4.4 Balance of Plant (BOP) Equipment	143
5	**TECHNICAL DESIGN ISSUES**	**143**
	5.1 Selecting and Sizing the Prime Mover	143
	5.2 Matching Electrical and Thermal Loads	144
	5.3 Dynamic Power and Thermal Matching	145
	5.4 Packaged Systems	146
6	**REGULATORY CONSIDERATIONS**	**147**
	6.1 Federal Regulations Related to Cogeneration	147
	6.2 Energy Policy Act of 2005	149
	6.3 Air Pollution Regulations	150
	6.4 Equipment Specific Regulations	151
	6.5 Water Quality and Solid Waste Disposal	153
	6.6 Permits and Certificates for Cogeneration	154
7	**ECONOMIC EVALUATIONS**	**154**
	7.1 Operating Costs of Current System	155
	7.2 Operating Costs of the Proposed Cogeneration System	157
	7.3 Economic Merit	157
8	**OWNERSHIP AND FINANCIAL ARRANGEMENTS**	**159**
	8.1 Overall Considerations	159
	8.2 Conventional Ownership and Operation (100 Percent Ownership)	161
	8.3 Partnership Arrangements	161
	8.4 Third-party Ownership	162
	8.5 Final Comments on Financial Aspects	162
9	**SUMMARY AND CONCLUSIONS**	**163**

1 INTRODUCTION

The term *cogeneration* refers to the combined production of electrical power and useful thermal energy by the sequential use of a fuel or fuels. The electrical power is produced by an electrical generator, which is most often powered by a prime mover such as a steam turbine, gas turbine, or reciprocating engine. Examples of useful thermal energy include hot exhaust gases, hot water, steam or chilled water.

Cogeneration is important because of the potential for monetary and energy savings, and emission reductions. Any facility that uses electrical power and has thermal energy needs is a candidate for cogeneration. Although many considerations are involved in determining if cogeneration is feasible for a particular facility, the basic consideration is if the savings on thermal energy costs are sufficient to justify the capital expenditures for a cogeneration system. Facilities that may be considered for cogeneration include those in the industrial, commercial, and institutional sectors.

The technology for cogeneration exists for a range of sizes: from less than 50 kW to over 100 MW. The major equipment requirements include a prime mover, electrical generator, electrical controls, heat recovery systems, and other typical power plant equipment. These components are well developed, and the procedures to integrate these components into cogeneration systems are well established.

In addition to the economic and technical considerations, the application of cogeneration systems involves an understanding of the governmental regulations and legislation on electrical power production and on environmental impacts. With respect to electrical power production, certain governmental regulations were passed during the late 1970s, which removed barriers and provided incentives to encourage cogeneration development. Finally, no cogeneration assessment would be complete without an understanding of the financial arrangements, contracts, and agreements that are possible.

The sections of this brief overview of cogeneration systems will include introductory comments, descriptions of basic systems and terminology, descriptions of prime movers and major equipment, some comments on technical designs, a summary of relevant regulations, descriptions of economic evaluations, and comments on financial and ownership aspects. Several references cover various aspects of cogeneration systems (e.g., Refs. 1–7).

1.1 History of Cogeneration

At the beginning of the twentieth century, electrical power generation was in its infancy. Most industrial facilities generated all their own electrical power, and often supplied power to nearby communities. They used the thermal energy that was available during the electrical power production to provide or supplement process or building heat. These industrial facilities, therefore, were the first *cogenerators.* The dominant prime mover at this time was the reciprocating steam engine, and the low-pressure exhaust steam was used for heating applications.

Between the early 1920s and the 1960s, the public electric utility industry grew rapidly because of increasing electrical power demands. Coincident with this rapid growth was a general reduction in the costs to produce the electrical power, mainly due to economies of scale, more-efficient technologies, and decreasing fuel costs. During this period, industry often abandoned its own electrical power generation, because of four factors:

1. Decreasing electrical rates charged by public utilities
2. Income tax regulations, which favored expenses instead of capital investments
3. Increasing costs of labor
4. The desire of industry to focus on their product rather than the side issue of electrical power generation.

Estimates are available that suggest that industrial cogenerated electrical power decreased from about 25 to 9 percent of the total electrical power generated in the country between the years of 1954 and 1976. Since about the mid-1980s, this percentage has been fairly constant at about 5 percent. For example, at the end of 1992, 5.1 percent of the total U.S. electrical capacity was due to cogeneration systems.

In late 1973 and again in 1979, America experienced major energy crises, which were largely a result of reduced petroleum imports. Between 1973 and 1983, the prices of fuels and electrical power increased by a factor of about five. Any facility purchasing electrical power began to consider (or reconsider) the economic savings associated with cogeneration. These considerations were facilitated by federal regulations that were enacted in 1978 to ease or remove barriers to cogeneration.

1.2 Constraints on Cogeneration

Although the arguments for cogeneration technology are persuasive, a number of obstacles may constrain the implementation of these systems:

- *High cost of capital investment.* Costs of cogeneration systems vary, depending on the size and the type of facility. For relatively large systems, these costs can be millions of dollars.
- *High cost of fuel.* The fuel cost can be the major operating expense of a cogeneration facility.
- *Low cost of electricity.* Despite the rate increases of recent years, the cost of electricity still remains low in many areas of the country and for certain sectors (such as for large industrial users due to the declining block rate structuring approach used by some utilities).

- *Environmental concerns.* The regulations on environmental emissions continue to impede the implementation of new power facilities. In some areas of the country (e.g., California), new power-plant construction has slowed or stopped for some periods of time.
- *Restricted revenue from electricity sales* The Federal Energy Regulatory Commission (FERC) has required utilities to purchase cogenerated electricity, minimizing this obstacle, but the utilities pay a rate on an *avoided cost* basis.
- *High back-up rates.* Electric utilities have traditionally charged high rates to provide stand-by power. The FERC has ruled that electric utilities must apply the theory of load diversity in a nondiscriminatory fashion to establish stand-by rates.

2 BASIC COGENERATION SYSTEMS

2.1 Topping Cycles

A cogeneration system may be classified as either a topping cycle system or a bottoming cycle system. Figure 1 is a schematic illustration of a topping-cycle system. As shown, a prime mover uses fuel to power an electrical generator to produce electricity. This electricity may be used completely on-site or may be tied into an electrical distribution network for sale to the local utility or other customers. The hot exhaust gases are directed to a heat recovery steam generator (HRSG)* to produce steam or hot water. This steam or hot water is

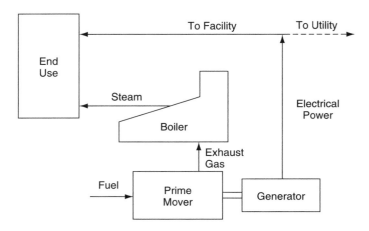

Figure 1   A schematic illustration of a cogeneration topping-cycle system.

---

*Many other terms for this boiler are common such as waste heat boiler (WHB), and heat recovery boiler (HRB).

used on-site for process or building heat. This cogeneration system is classified as a topping-cycle because the electrical power is generated first at the higher (*top*) temperatures associated with the fuel combustion process, and then, the rejected or exhausted energy is used to produce useful thermal energy. The majority of cogeneration systems are based on topping cycles.

2.2 Bottoming Cycles

The other classification of cogeneration systems is bottoming-cycle systems. Figure 2 is a schematic illustration of a bottoming cycle system. As shown, the high-temperature combustion gases are used first in a high-temperature thermal process (such as high-temperature metal treatment), and then, the lower-temperature gases are used in a special low-temperature cycle to produce electrical power. After the energy is removed at the high temperatures, the energy available at the *bottom* or lower temperatures is then used to produce electrical power.*

Bottoming-cycle cogeneration systems have fewer applications than topping-cycle systems, and must compete with waste-heat recovery systems such as feedwater heaters, recuperators, and process heat exchangers. One of the

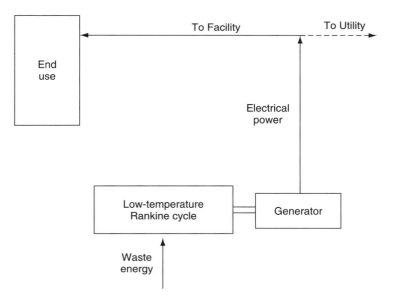

Figure 2 A schematic illustration of a cogeneration bottoming-cycle system.

---

*Other definitions of bottoming cycles are common. These other definitions often include any second use of the energy. For example, some authors refer to the steam turbine in a combined cycle as using a bottoming cycle. The more precise thermodynamic definition employed here is preferred, although fewer applications meet this definition.

difficulties with bottoming-cycle systems is the low-temperature electrical power-producing cycle. One example, depicted in Figure 2, is a low-temperature Rankine cycle. The low-temperature Rankine cycle is a power cycle similar to the conventional steam Rankine cycle, but a special fluid such as an organic substance (like a refrigerant) is used in place of water. This fluid vaporizes at a lower temperature compared to water, and therefore, this cycle is able to utilize the low temperature energy. These cycles are generally much less efficient than conventional power cycles, often involve special equipment, and use more expensive working fluids.

2.3 Combined Cycles

One power-plant configuration, based on a form of a topping cycle and widely used in industry and by electrical utilities, is known as a combined cycle. Figure 3 is a schematic illustration of a possible combined-cycle cogeneration system. In this example, a gas turbine generates electrical power, and the exhaust gas is ducted to an unfired heat recovery boiler. The produced steam then drives a steam turbine, which produces additional electrical power. The exhaust steam from the steam turbine is at a high enough pressure and temperature to supply thermal energy for process or building heat. In this example, the steam is then condensed and pumped back into the boiler. For such a combined cycle gas turbine (CCGT) power plant to qualify as a cogeneration application, some steam

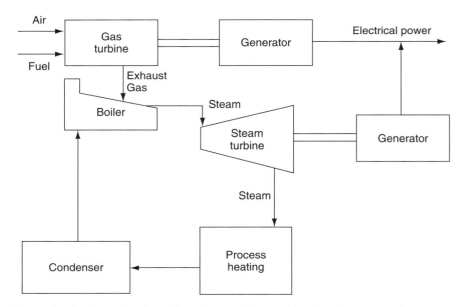

Figure 3   A schematic illustration of a possible combined-cycle cogeneration system.

would need to be used to satisfy a thermal requirement. If no thermal commodity is produced and used, the facility could not be considered a cogeneration system.

As might be expected, combined cycles have high power-to-heat ratios and high electrical efficiencies. Current designs may have electrical efficiencies as high as 60 percent, depending on the equipment, location, and details of the specific application. These current designs for combined cycle plants result in the gas turbine power to be 1.5 to 3.5 times the power obtained from the steam turbine. These plants are most often base-load systems operating more than 6,000 hours per year. More details on gas turbines and steam turbines are provided in the following sections on the prime movers.

2.4   Applications of Cogeneration Systems

Cogeneration systems may involve different types of equipment, and may be designed to satisfy specific needs at individual sites. However, many sites have similar needs and packaged (pre-engineered) cogeneration systems may satisfy these needs and are less expensive than custom engineered systems.

Cogeneration systems are found in all economic sectors of the world. For convenience, cogeneration systems are often grouped into one of three sectors: (1) industrial, (2) institutional, or (3) commercial. The types and sizes of the cogeneration systems in these sectors overlap to varying degrees, but nonetheless, these sectors are convenient for describing various applications of cogeneration. Examples of successful applications are often found for universities (and other similar campuses), hospitals, other medical facilities, military bases, industrial sites, laundries, hotels, and airports.

3   DESCRIPTIONS OF PRIME MOVERS

Cogeneration systems consist of several major pieces of equipment and many smaller components. This section will describe the prime movers, and the following section will describe the other major equipment (electrical equipment, heat recovery devices, absorption chillers, and balance of plant equipment).

Prime movers include those devices that convert fuel energy into rotating shaft power to drive electrical generators. The prime movers that are used most often in cogeneration systems are steam turbines, gas turbines, and reciprocating engines. Important distinctions between the prime movers are the fuels that they may use, their combustion processes, their pollutant emissions, their overall thermal efficiency, and the type, amount, and temperature of their rejected thermal energy. In cogeneration applications, a significant parameter for each type of prime mover is the ratio of the rate of supplied thermal energy and the output power. This ratio is called the *heat-to-power ratio*. Knowing the value of the heat-to-power ratio assists in matching a particular prime mover to a particular application. This matching is discussed in a subsequent section.

3.1 Steam Turbines

Steam turbines are widely used in power plants throughout industry and electric utilities. Steam turbines use high-pressure, high-temperature steam from a boiler. The steam flows through the turbine, forcing the turbine wheel to rotate. The steam exits the turbine at a lower pressure and temperature. A major advantage of the steam turbine relative to reciprocating engines and gas turbines is that the combustion occurs externally in a separate device (boiler). This allows a wide range of fuels to be used, including solid fuels such as coal or solid waste materials. The turbine's exit steam, of course, can be used for thermal heating or to supply the energy to an absorption chiller.

Steam turbines are available in a multitude of configurations and sizes. A major distinction is whether the machine is a condensing or noncondensing (back-pressure) steam turbine. Condensing steam turbines are steam turbines designed so that the steam exits at a low pressure (less than atmospheric) such that the steam may be condensed in a condenser at near ambient temperatures. Condensing steam turbines provide the maximum electrical output, and hence, are most often used by central plants and electric utilities. Since the exiting steam possesses little available energy, applications of condensing steam turbines for cogeneration would require the extraction of steam prior to the exhaust.

Noncondensing steam turbines are those steam turbines that are designed so that the exiting steam is at a pressure above atmospheric. The exiting steam possesses sufficient energy to provide process or building heat. Either type of steam turbine may be equipped with one or more extraction ports so that a portion of the steam may be extracted from the steam turbine at pressures between the inlet and exit pressures. This extracted steam may be used for heating or thermal processes which require steam at higher temperatures and pressures than that which is available from the exiting steam.

Noncondensing steam turbines are available in a wide range of outputs beginning at about 50 kW and increasing to over 100 MW. Inlet steam pressures typically range from 150 to 2000 psig, and inlet temperatures range from $500°$ to $1,100°$F. Depending on the specific design and application, the heat-to-power ratio for steam turbines could range from 4 to over 10. The thermal efficiency typically increases with size (or power level). Although the major source of thermal energy is the exit or extracted steam, the boiler exhaust may be a possible secondary source of thermal energy in some cases.

3.2 Gas Turbines

As with steam turbines, stationary gas turbines are major machines in many power plants. Stationary gas turbines share many of the same components with the familiar aircraft gas turbines. In fact, both stationary (or industrial) and aircraft (or aero-derivative) gas turbines are used in cogeneration systems. The major components of a gas turbine are the air compressor, the combustor, and the turbine. A significant fraction of the turbine power is used internally to drive the compressor. This brief description will highlight the important characteristics of gas turbines as applied to cogeneration.

Many configurations, designs, and sizes of gas turbines are available. The simple-cycle gas turbine uses no external techniques such as regeneration to improve its efficiency. The thermal efficiency of simple cycle gas turbines may be increased, therefore, by the use of several external techniques, but the designs and configurations become more complex. Many of these modifications to the simple cycle gas turbine are directed at using the energy in the exhaust gases to increase the electrical output and efficiency. Of course, such modifications will decrease the available energy in the exhaust. For some cogeneration applications, therefore, the most efficient gas turbine may not always be the appropriate choice.

The single shaft, single turbine described above is the configuration of the simple cycle gas turbine. Other configurations are available. Gas turbines may be designed with two or more turbines. This permits one turbine to be designed for high rotating speeds to drive the compressor, and a second (mechanically uncoupled) turbine to be operated at generator speeds. This flexibility permits a more overall efficient design. These gas turbines are known as two- or three-shaft machines. The multiple shaft machines are more complex, and hence, more costly than the simple single-shaft machines.

A gas turbine may also be equipped with regeneration, intercooling, and reheating. *Regeneration* (also known as *Recuperation*) is the process of using exhaust gas energy to heat the air from the compressor before the air enters the combustor. This lowers the fuel consumption of the gas turbine for the same combustor outlet gas temperature, but regeneration will reduce the energy (temperature) of the exhaust gases for cogeneration applications. *Intercooling* is the process of cooling the partially compressed air. Intercooling would normally be installed between stages of a gas turbine which used two or more compressor stages. The use of intercooling reduces the required compressor power, and therefore, increases the turbine output power. *Reheating* is the process of providing other combustors after the main combustor. Reheating is especially effective where two or more turbines are used. The gases may be reheated between the multiple turbines. Other modifications and variations of gas turbines are available, but these are the most common.

A variety of combustor designs are used in different gas turbines. These designs are aimed at providing stable combustion, long life, and low emissions. Typically, the combustor has a primary zone that operates near stoichiometric, and then the product gases are diluted with additional air. This dilution is necessary to reduce the gas temperatures to acceptable levels for the turbine blades. The final product gas mixture will represent a high air-fuel mass ratio (for some cases, the total air mass flow rate may be on the order of 100 times the fuel mass flow rate). In other words, the gas turbine operates with high levels of overall excess air. Due to the large amount of excess air used in the combustion process of gas turbines, the exiting exhaust gas contains a relatively high concentration of nitrogen and oxygen. Hence, the gas turbine exhaust may be characterized as mostly heated air, and is nearly ideal for process or heating purposes.

Gas turbines may use liquid fuels such as jet fuel or kerosene, or they may use gaseous fuels such as natural gas or propane. The highest performance is possible with liquid fuels, but the lowest emissions have been reported for natural gas operation.

3.3   Reciprocating Engines

A third category of prime movers for cogeneration systems is internal combustion (IC), reciprocating engines.* These engines are available in several forms. Probably the most familiar form of the reciprocating engine is the typical spark-ignited, gasoline engine used in automobiles. For cogeneration applications, the spark-ignited gasoline engine must be converted to operate in a stationary, continuous mode with fuels such as natural gas. Such engines are typically for small cogeneration systems with less than about 100 kW of electrical output. One major group of reciprocating engines for mid- to large-sized cogeneration systems are stationary diesel engines operating with either diesel fuel, or in a dual-fuel mode with natural gas. Another large number of reciprocating engines for cogeneration systems are stationary gas engines using natural gas fuel and spark ignition. All of these engines share some common characteristics for cogeneration applications, and have some distinctive features as well.

Power ratings for reciprocating engines are similar to those for gas turbines in that both continuous and intermittent duty cycle ratings are provided. As with the gas turbines, these power ratings are provided for a set of standard conditions for ambient temperature and pressure, and elevation. The standard power ratings need to be adjusted for the local conditions at the site of the installation. Reciprocating engines are not as adversely affected by high inlet air temperatures, as are gas turbines. Furthermore, many larger reciprocating engines are equipped with turbochargers and after-coolers, which minimize the effects of inlet air conditions. For cogeneration applications, reciprocating engines are available in many power levels and designs. These power levels range from less than 50 kW to over 60 MW for single engines. Some manufacturers even offer "mini" cogeneration systems with outputs as low as 6 kW.

The portion of the fuel energy that is not converted into mechanical power ultimately is rejected to the surroundings. This energy is rejected to the cooling water and lubricating oil, and to the surroundings by radiation from the engine block and by the hot exhaust gases. The fraction of energy rejected in these different manners depends on the engine design and operating conditions. As an example, if 35 percent of the fuel energy is converted to shaft power output, then

---

*Although rotary engines could be used in cogeneration systems, at this time no significant applications are known. The remaining discussion will focus on reciprocating engines.

30 percent of the fuel energy may be rejected to the cooling liquid, 27 percent may be rejected with the exhaust gas, and 8 percent may be rejected as radiation and miscellaneous other energy rejections.

For those reciprocating, internal combustion engines that are liquid cooled (the majority of the engines considered here), the cooling liquid is a secondary source of thermal energy. Although not at the high temperatures of exhaust gas, this energy can be used to produce hot water. Several designs are available for recovering the energy in the cooling liquid. These designs use one or more direct or indirect heat exchangers to generate the hot water or low pressure steam. Liquid-to-liquid heat exchangers can have high efficiencies, and most of this energy is recoverable (but at relatively low temperatures). Other sources of energy from a reciprocating engine are sometimes possible to recover, such as from oil-coolers and turbocharger after-coolers. This energy is usually at temperatures below 160°F, and would only be practical to recover for low-temperature requirements.

Another benefit of the reciprocating engine is that the maintenance and repair is less specialized than for gas turbines. However, the maintenance may be more frequent and more costly.

3.4   Other Possible Prime Movers

Although most cogeneration systems are based on the prime movers, some other possibilities exist. Some cogeneration systems are based on fuel cells. Fuel cells generally use hydrogen to produce electricity. During this conversion process, thermal energy must be removed. This energy can be captured and used to produce hot water. Such a system would be a cogeneration plant.

Microturbines are also used in cogeneration systems. Although actually a subclassification of gas turbines, since they represent a relatively new technology, they are often described as a separate category of prime mover. The generic description for gas turbines would apply in general to microturbines. As small as a refrigerator, a microturbine may produce something on the order of 25 to 500 kW of electricity. Thermal energy in the exhaust is generally used to produce hot water.

4   DESCRIPTION OF OTHER EQUIPMENT AND COMPONENTS

In addition to the prime mover, cogeneration systems consist of several major pieces of equipment and many smaller components: (1) electrical equipment, (2) heat recovery devices, (3) absorption chillers, and (4) balance of plant equipment.

4.1   Electrical Equipment

The electrical equipment for cogeneration systems includes electrical generators, transformers, switching components, circuit breakers, relays, electric meters, controls, transmission lines, and related equipment. In addition to the equipment that supports electrical production, cogeneration systems may need equipment to interconnect with an electric utility to operate in parallel for obtaining supplementary power, as back-up (emergency) power, or for electrical sales to the utility.

The electric generator is a device for converting the rotating mechanical energy of a prime mover to electrical energy. The basic principle for this process, known as the *Faraday effect,* is that when an electrically conductive material such as a wire moves across a magnetic field, an electric current is produced in the wire. This can be accomplished in a variety of ways, and therefore, there are several types of electric generators. The frequency of the generator's output depends on the rotational speed of the assembly.

Most often the manufacturer of the prime mover will provide the prime mover and generator as an integrated, packaged assembly (called a *gen-set*). Performance characteristics of generators include power rating, efficiency, voltage, power factor, and current ratings. Each of these performance characteristics must be considered when selecting the proper generator for a given application. Electric generators may have conversion efficiencies of between about 50 to 98 percent, and, in general, the efficiency increases with increases in generator size (power level). Only the largest electric generators (say, on the order of 100 MW) attain efficiencies of 98 percent.

4.2   Heat-Recovery Equipment

The primary heat-recovery equipment used in cogeneration systems includes several types of steam and hot-water production facilities. In addition, absorption chillers could be considered in this section, but for organizational reasons, absorption chillers will be discussed in the following subsection.

Several configurations of heat-recovery devices are available. As already mentioned, these devices may be referred to as *heat-recovery steam generators,* or HRSGs. HRSGs are often divided into the following categories: (1) unfired, (2) partially fired, and (3) fully fired. An unfired HRSG is essentially a convective heat exchanger. A partially fired HRSG may include a *duct burner,* which often uses a natural gas burner upstream of the HRSG to increase the exhaust gas temperature. A fully fired HRSG is basically a boiler that simply uses the exhaust gas as preheated air. Figure 4 is a schematic of one configuration of an unfired HRSG. As shown in this schematic, gas turbine exhaust flows up through the device and exits at the top. Energy from the exhaust gas is used to heat and vaporize the water, and to superheat the steam.

Figure 5 shows the water/steam and exhaust gas temperatures for the three sections of a typical unfired HRSG: economizer, evaporator, and superheater. The top line shows the exhaust gas temperature decreasing from left to right as energy is removed from the gas to heat the water. The lower line represents the water heating up from right to left in the diagram. The lower-temperature exhaust

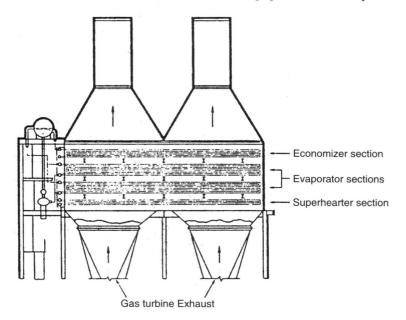

Figure 4 Schematic of an unfired heat-recovery steam generator (HRSG).

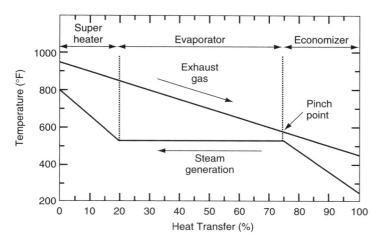

Figure 5 Temperature as a function of heat transfer (or position) for the three sections (economizer, evaporator, and superheater) of a typical unfired HRSG.

is used to preheat the water to saturation conditions in the economizer. The intermediate-temperature exhaust is used to vaporize (or boil) the water to form saturated steam. Finally, the highest-temperature exhaust is used to superheat the steam.

The temperature difference between the exhaust gas and the water where the water first starts to vaporize is referred to as the pinch point temperature difference. This is the smallest temperature difference in the HRSG and may limit the overall performance of the heat recovery device. On one hand since the rate of heat transfer is proportional to the temperature difference, the greater this difference the greater the heat transfer rate. On the other hand, as this temperature difference increases the steam flow rate must decrease and less of the exhaust gas energy will be utilized. To use smaller temperature differences and maintain higher heat transfer rates, larger heat-exchanger surfaces are required. Larger heat-transfer surface areas result in higher capital costs. These, then, are the types of trade-offs that must be decided when incorporating a heat recovery device into a cogeneration system design.

4.3   Absorption Chillers

Absorption chillers may use the thermal energy from cogeneration systems to provide cooling for a facility. Absorption chillers use special fluids and a unique thermodynamic cycle, which produces low temperatures (for the cooling) without the requirement of a vapor compressor, which is used in mechanical chillers. Instead of the vapor compressor, an absorption chiller uses liquid pumps and energy from hot water, steam, or exhaust gas.

For cogeneration applications, the important feature of absorption chillers is that they use relatively low-temperature energy available directly or indirectly from the prime mover and produce chilled water for cooling. The use of absorption chillers is particularly advantageous for locations where space and water heating loads are minimal during a good part of the year. For these situations, the thermal output of a cogeneration system can be used for heating during the colder part of the year and, using an absorption chiller, for cooling during the warmer part of the year. Furthermore, by not using electric chillers, the electric loads are more constant throughout the year. In warm climates, absorption chillers are often an important, if not an essential, aspect of technically and economically successful cogeneration systems.

Some absorption chillers are designed as indirect-fired units using hot-water or steam. As examples of typical numbers, a single-stage unit could use steam at $250°F$ to produce a ton of cooling for every 18 pounds of steam flow per hour. A dual-stage unit, would need $365°F$ steam to produce a ton of cooling for every 10 pounds of steam flow per hour. If hot water is available, a ton of cooling could be produced for every 220 pounds of $190°F$ hot water per hour.

Other absorption chillers use the exhaust gas directly and are called *direct-fired units.* Direct-fired absorption chillers are particularly advantageous when a steam or hot water system does not exist. For a direct-fired absorption chiller, the exhaust gas temperature needs to be $550°$ to $1,000°F$. The higher the exhaust temperature, the less energy (or exhaust gas flow) is needed per ton of cooling.

For example, for 1,000°F exhaust gas, a ton of cooling requires 77 pounds per hour of flow whereas for 550°F exhaust gas a ton of cooling requires 313 pounds per hour of flow.

4.4 Balance of Plant (BOP) Equipment

Balance of plant equipment includes those components not explicitly described already. The BOP equipment for cogeneration systems is similar to that for conventional power plants. This includes other controls, emergency devices, exhaust systems and stacks, natural gas compressors, any thermal energy storage equipment, water treatment devices, concrete bases or pads, fuel supply system components, any necessary building modifications, other piping and fittings, mechanical system interfaces, condensers, cooling systems, feedwater tanks, deaerators, feedwater pumps, other pumps, flue gas bypass valves, dampers and ducts, and other such equipment.

5 TECHNICAL DESIGN ISSUES

5.1 Selecting and Sizing the Prime Mover

The selection of a prime mover for a cogeneration system involves the consideration of a variety of technical and nontechnical issues. Technical issues, which often dominate the selection process, include the operating mode or modes of the facility, the required heat-to-power ratio of the facility, the overall power level, and any special site considerations (e.g., low noise). Other issues, which may play a role in the selection process, include the desire to match existing equipment, and to utilize the skills of existing plant personnel. Of course, the final decision is often dominated by the economics.

Steam turbines and boilers usually are selected for a cogeneration system if the fuel of choice is coal or another solid fuel. For certain situations, a steam turbine system may be selected even for a liquid or gaseous fuel. Also, steam turbines and boilers would be selected if a high heat-to-power ratio is needed. Steam turbines also may be selected for a cogeneration system in certain specialized cases. For example, a large pressure-reduction valve in an existing steam system could be replaced with a steam turbine and, thereby, provide electrical power and thermal energy. In other applications, steam turbines are selected to be used in conjunction with a gas turbine in a combined-cycle power plant to increase the electrical power output. Combined-cycle gas turbine power plants for cogeneration system applications were described in an earlier section.

Gas turbines are selected for many cogeneration systems where the required heat-to-power ratio and the electrical power need are high. Also, gas turbines are the prime mover of choice where minimal vibration or low weight to power (such as for a roof installation) is required. Reciprocating engines are selected where the heat-to-power ratio is modest, the temperature level of the thermal energy is low, and the highest electrical efficiency is necessary for the economics. Usually, for the smaller systems, reciprocating engines will result in the most favorable economics. Additionally, reciprocating engines may be selected if the plant personnel are more suited to the operation and maintenance of these engines.

Selecting the appropriate size prime mover involves identifying the most economic cogeneration operating mode. This is accomplished by first obtaining the electrical and thermal energy requirements of the facility. Next, various operating modes are considered to satisfy these loads. By conducting a comprehensive economic analysis, the most economic operating mode and prime mover size can be identified. The process of matching the prime mover and the loads is described next.

5.2 Matching Electrical and Thermal Loads

To properly select the size and operating mode of the prime mover, the electric and thermal loads of the facility need to be obtained. For the most thorough matching, these loads are needed on an hourly, daily, monthly, and yearly basis. As an example, Figure 6 shows the month totals for the electrical and thermal loads for a hypothetical facility. For the summer months (numbers 5–9), the heating loads are minimum, and then for the winter months, the heating loads are higher. The electrical loads are highest for the summer months reflecting the use of air conditioning. In addition, this figure shows dashed lines, which represent the *base loads* for the electrical and heating loads, respectively. The base loads are the minimum loads during the year, and form a floor or base for the total loads. Often, a cogeneration system may be sized so as to

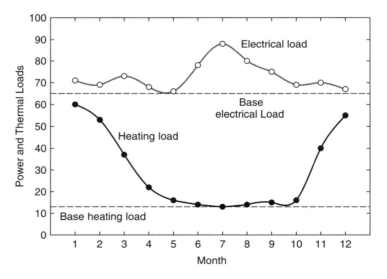

Figure 6 An example of monthly electrical and thermal loads.

provide only the base loads. In this case, auxiliary boilers would provide the additional heating needed during the days where the heating needs exceeded the base amount. Similarly, electrical power would need to be purchased to supplement the base power provided by the cogeneration system.

The possible overall operating modes for a cogeneration power plant are often categorized into one of three classes.

1. The plant may operate as a *base-load system* with little or no variation in power output. Base load plants operate in excess of 6,000 hours per year. Power needs above the base load are typically provided by interconnections to a local utility or by an auxiliary power plant.

2. The plant may operate as an *intermediate system* for 3,000 to 4,000 hours per year. These systems are less likely than base-load systems, but if the economics are positive, may have application for facilities that are not continuously operated, such as some commercial enterprises.

3. The plant may use a *peaking system,* which operates for only 1,000 hours or less per year. Utility plants often use peaking systems to provide peaking power during periods of high electrical use. For cogeneration applications, peaking units may be economical where the costs of the electricity above a certain level is unusually high. These units are sometimes referred to as *peak shaving systems.*

5.3 Dynamic Power and Thermal Matching

In addition to selecting and sizing the prime mover for the average loads, consideration must be given to dynamic operation of the cogeneration system. Dynamic operation refers to the necessity of satisfying the *minute-by-minute* electrical power and thermal needs of a facility.

The options for electrical power modulation include operating the prime mover at part load as needed. This is essentially *load following.* The disadvantage to this approach is the lower efficiencies at part load. The second option is to use multiple prime movers. This option allows the prime movers to be operated near full load more often. When the power requirements increase, one or more additional prime movers would be activated. One disadvantage of this option is that the economics may be less attractive since multiple units are generally more expensive than a single, larger unit. Another disadvantage is the additional wear and deterioration of the prime mover due to the more frequent starting and stopping of individual prime movers. A third option (and often the most common) is to use the utility power to make up any power needed in excess of what the prime mover can supply. The disadvantage is the utility charges for electrical power, and the dependence on the utility.

The options for providing the dynamic thermal loads of a facility include the use of one or more supplementary boilers. This is the most common form of thermal modulation, and simply requires the supplementary boiler or boilers to *follow* the thermal loads, and provide the thermal requirements in excess of what can be supplied by the cogeneration system. The next option is heat dumping, which is simply discharging the thermal energy not needed. This is obviously not attractive from an energy conservation or economic perspective, but may be acceptable as a short-term solution in cases where future thermal needs are expected to increase. The last option is to utilize some form of thermal storage. In this case, thermal energy is stored during periods of excessive production of thermal energy. The storage options are varied, but a common technique is the use of hot water tanks.

The overall topic of dynamic matching of the electrical power and thermal needs is complex, and is central to the overall issue of selecting and sizing the prime movers for a given facility. Many scenarios need to be explored, and the resulting economics examined. Once all the issues are fully explored, decisions may be made on what technical design makes the most sense for a given facility.

5.4 Packaged Systems

In general, facilities with low electrical power needs cannot utilize customized cogeneration systems because of the relatively high initial costs that are associated with any system. These initial costs include at least a portion of the costs related to the initial design, engineering, and related development and installation matters. Also, smaller facilities often do not have the specialized staff available to develop and operate complex power plants.

To solve some of these problems, pre-engineered, factory-assembled, *packaged* cogeneration systems have been developed. The major advantage of packaged cogeneration systems is the fact that the initial engineering, design, and development costs can be spread over many units, which reduces the capital cost (per kW) for these systems. Other advantages of packaged cogeneration systems include factory assembly and testing of the complete system. If there are any problems, they can be fixed while the system is still at the manufacturer's plant. The standard design and reduced installation time result in short overall implementation times. In some cases, a packaged cogeneration system could be operational within a few months after the order is received. This short implementation time reduces the project's uncertainty, which eases making decisions and securing financing.

Another advantage of packaged cogeneration systems is the fact that the customer only interacts with one manufacturer. In some cases, the packaged cogeneration system manufacturer will serve as project engineer, taking the project from initial design to installation to operation. The customer often may decide to purchase a *turn-key system*. This provides the customer with little uncertainty, and places the burden of successful project completion on the manufacturer. Also, the manufacturer of a packaged cogeneration system will have experience interacting with regulating boards, financing concerns, and utilities, and may assist the customer in these interactions.

The major disadvantage of packaged cogeneration systems is that the system is not customized for a specific facility. This may mean some compromise and lack of complete optimization. Specialized configurations may not be available. Also, beyond a certain size, packaged cogeneration systems are simply not offered, and a customized unit is the only alternative.

6 REGULATORY CONSIDERATIONS

This section includes a brief overview of the relevant federal regulations on electric power generation by cogenerators and the related environmental constraints. Specifically, this section contains the following sections: federal regulations on power generation related to cogeneration, air pollution regulations, water and waste pollution regulations, and permitting and certificates for cogeneration.

6.1 Federal Regulations Related to Cogeneration

The passage of the Public Utility Regulatory Policies Act (PURPA) in 1978 helped cogeneration to become a much more attractive option for electric power generation for a variety of facilities. Prior to the passage of PURPA, there were three common barriers to cogeneration:

1. There was no general requirement by electric utilities to purchase electric power from cogenerators.
2. There was discriminatory back-up power for cogenerators (i.e., prices to cogenerators for back-up power were higher than the prices offered to other customers).
3. Cogenerators feared that they might become subject to the same state and federal regulations pertaining to electric utilities.

The passage of PURPA helped to remove these obstacles to development of cogeneration facilities.

Although PURPA was passed in 1978, it was not until 1980 that the Federal Energy Regulatory Commission (FERC) issued its final orders on PURPA. In the National Energy Act, FERC was designated as the regulatory agency for implementation of PURPA. The regulations dealing with PURPA are contained in Part 292 of the FERC regulations. Sections 201 and 210 are the two primary sections relevant to small power production and cogeneration.

Section 201 contains definitions of cogeneration and sets annual efficiency standards for new topping cycle* cogeneration facilities, which use oil or

---

*Since bottoming cycle cogeneration facilities do not use fuel for the primary production of electrical power, these facilities are only regulated when they use oil or natural gas for supplemental

natural gas.[†] For a cogenerating facility to qualify for the privileges and exclusions specified in PURPA, the facility must meet these legislated standards. These standards define a legislated or artificial *efficiency* that facilities must equal or exceed to be considered a *qualified facility* (QF). A qualified facility is eligible to use the provisions outlined in PURPA regarding nonutility electric power generation. This legislated *efficiency* is defined as

$$\eta_{\text{PURPA}} = \frac{\left(P + \frac{1}{2}T\right)}{F} \tag{1}$$

where P is the electrical energy output, T is the used thermal energy, and F is the fuel energy used (all items in consistent units). The "1/2" in this relation helps to encourage systems to have significant electrical power to obtain acceptable PURPA efficiencies.

Table 1 lists the standards for the PURPA efficiencies. These standards state that a facility must produce at least 5 percent of the site energy in the form of useful thermal energy. For cases where the useful thermal energy percentage is between 5 percent and 15 percent, the facility must have a PURPA efficiency of at least 45.0 percent. If the thermal fraction is greater than 15 percent, the facility meets the standards if it has an efficiency of at least 42.5 percent. Values for the thermal percentage and the PURPA efficiency are based on projected or estimated annual operations.

The purpose of introducing the artificial standards was to ensure that useful thermal energy was produced on site in sufficient quantities to make the cogenerator more efficient than the electric utility. Any facility that meets or exceeds the required efficiencies will be more efficient than any combination of techniques producing electrical power and thermal energy separately. Section 201 also put limitations on cogenerator ownership, that is, electric utilities could not own a majority share of a cogeneration facility, nor could any utility holding company,

Table 1   Required Efficiency Standards for Qualified Facilities (QF)

If the Useful Thermal Energy Fraction Is:	The Required η_{PURPA} Must Be:
≥5.0%	≥45.0%
≥15.0%	≥42.5%

firing. The standard states that, during any calendar year, the useful power output of the bottoming cycle cogeneration facility must equal or exceed 45 percent of the energy input if natural gas or oil is used in the supplementary firing. The fuels that are used first in the thermal process prior to the bottoming cycle cogeneration facility are not taken into account for satisfying PURPA requirements.

[†]For topping cycle cogeneration facilities using energy sources other than oil or natural gas (or facilities installed before March 13, 1980), no minimum has been set for efficiency.

nor a combination thereof. These ownership restrictions were removed in the Energy Act of 2005 (described in the next section).

Section 210 defines the procedures for obtaining QF status. An owner or operator of a generating facility may obtain QF status by either submitting a self-certification or applying for and obtaining a Commission certification of QF status. The choice of whether to certify a facility through a self-certification or Commission certification is up to the applicant. In some instances, negotiations with a lender or utility purchaser may proceed more smoothly if the facility has been certified by the Commission.

Section 210 of the PURPA regulations specifically addressed these three major obstacles to developing cogeneration facilities. The principal issues in Section 210 include the following legal obligations of the electric utility toward the cogenerator:

1. Obligation to purchase cogenerated energy and capacity from QFs
2. Obligation to sell energy and capacity to QFs
3. Obligation to interconnect
4. Obligation to provide access to transmission grid to wheel to another electric utility
5. Obligation to operate in parallel with QFs
6. Obligation to provide supplementary power, back-up power, maintenance power, and interruptible power

Section 210 also exempted QFs from utility status, and established a cost basis for purchase of the power from QFs. FERC specified that the price paid to the QF must be determined both on the basis of the utility's avoided cost for producing that energy and, if applicable, on the capacity deferred as a result of the QF power (i.e., the cost savings from not having to build a new power plant). Other factors, such as QF power dispatchability, reliability, and cooperation in scheduling planned outages, could also be figured into the price paid to the QFs by the electric utilities. The state public utility commissions were responsible for determining the value of these avoided cost rates.

6.2   Energy Policy Act of 2005

The Energy Policy Act (EPAct) of 2005 contained several items that specifically concerned cogeneration. In particular, EPAct eliminated the ownership limitations on qualifying facilities, which were part of the original PURPA. This means, for example, that utilities are allowed to own up to 100 percent of a cogeneration plant (while the original PURPA restricted utilities to less than 50 percent ownership). For qualifying facilities that have nondiscriminatory access to other sources of electric power (such as from wholesale markets), EPAct relaxed the requirements that utilities must purchase cogenerated power and must sell electric energy to qualifying facilities. EPAct also contains new language, which encourages the thermal output of cogeneration plants to be productive and useful (and to avoid situations where no real need existed for the thermal energy).

6.3   Air Pollution Regulations

Legislation to limit pollutant emissions has a long history. In the United States, the first major national legislation was the Air Pollution Control Act in 1955. This legislation was motivated largely by the recognition of the air quality concerns in California, and particularly, in the Los Angeles area. This act was narrow in scope, and provided no specific limitations on pollutant emissions.

The current era of air pollution regulation was started with the Clean Air Act (CAA) of 1963. The CAA was considered the first major modern environmental law established by the United States congress, and set the ground work for the current regulation format. The original CAA has been revised by Congress six times since it became law, with major amendments in 1967, 1970, 1977, and 1990. These amendments are referred to as the Clean Air Act Amendments (CAAA). Each of these subsequent amendments continued to increase the strength of the original law by lowering the acceptable levels of emissions. The CAA and the subsequent CAAAs apply to a wide range of applications, and they have a significant impact on the design and operation of power generation facilities.

The first major revision to the CAA occurred in 1967, which established the National Ambient Air Quality Standards (NAAQS) and required State Implementation Plans (SIPs) to verify compliance on the state level. In addition, the amendment established Air Quality Control Regions (AQCR) to interconnect different states into larger regional areas, since air pollution is not restricted to state boundaries. The NAAQS established maximum safe levels for the different pollutants designated under the original CAA. Although these standards did not have a direct effect on power-generation facilities, it was the precursor to several minor amendments between 1970 to 1975 that established federal emission limits for specific equipment. These amendments were used as a basis for the new source performance standards (NSPS), which apply to both new and modified stationary sources. These stationary sources include not only power generation facilities, but also a larger number of industrial operations such as municipal waste combustors, sulfuric acid production units, and grain elevators. These intermediate amendments also established regulations for mobile sources of air pollution (e.g., automobiles) and also for hazardous air pollutants. The NSPS regulation only takes into account the pollutant emissions from the source regardless of the surrounding environment. This required facilities to control or reduce pollutant emissions even though the ambient air pollution levels may be significantly lower then the NAAQS amounts.

One of the major amendments to the CAA was passed in 1977 and added two additional regulatory programs: the nonattainment (NA) program and the prevention of significant deterioration (PSD) program. These two programs addressed the issue of meeting the previously established ambient air-quality standards for the air-quality control regions. The nonattainment program applied to the regions that failed to meet the ambient air standards and the PSD program was designed to preserve and protect air quality in regions surpassing the national standards. The PSD program set the allowable pollutant levels lower than the nonattainment areas to prevent previously clean or unpolluted areas from becoming polluted. The PSD program also required the large size facilities (facilities producing over 100 tons/yr of a controlled pollutant) to include the *best available control technology* (BACT) to reduce emission levels below that required by the NSPS. Under the NA program, new facilities in nonattainment zones must be equipped with controls to assure the *lowest achievable emissions rate* (LAER). The facility must also show that some other source of pollutants must be reduced or eliminated to provide a net increase in air quality. The NA program also affects existing facilities by requiring timely reduction of emissions using *reasonable available control technology* (RACT). The difference between BACT, RACT, and LAER is that the first two include economic considerations when determining the required control equipment, while LAER is based on the most advanced equipment to achieve the lowest emissions without regard to cost. The addition of these two programs significantly increased the complexity of determining the emission limits for new and modified power-generation facilities.

The regulation of emission levels for power-generation facilities is determined by several different agencies at the federal, state, and local levels. State and local agencies can require lower emission levels, but can never require higher levels than the federal requirements. The Clean Air Act also allowed state and local authorities to reduce the acceptable levels to meet specific local air-quality standards. For example, the Los Angeles basin area is regulated by the South Coast Air Quality Management District (SCAQMD) and has stricter standards than those set by the federal government.

6.4 Equipment Specific Regulations

In addition to these requirements, regulations exist for specific equipment. In terms of power generation, the specific regulated equipment includes stationary gas turbines and boilers (steam systems). Stationary reciprocating engines are not specifically covered by NSPS since they have a much smaller impact on the overall generation capacity. The emissions from reciprocating engines will, of course, need to satisfy any local, state, or federal limits at the installation site. They are regulated extensively for mobile and some off-highway applications.

Stationary gas turbines have become one of the prominent types of power generation facilities. As a result of this growth, the Code of Federal Regulations (CFR) includes specific requirements for the emission levels of gas turbines. The NO_X emission limits are divided into two levels, depending on the energy input rate of the plant, and are increased if the fuel contains nitrogen. For natural gas–fired units, neither fuel-bound nitrogen nor sulfur dioxide exists in significant quantities. Also, the NSPS limits do not cover particulates or unburned hydrocarbon emissions for gas turbines, since neither exists in significant amounts.

Steam-based Rankine-cycle electric-power-generation facilities are also covered under the NSPS regulations. The allowable emissions levels vary based on the type of fuel used, with substantially lower limits for liquid and gaseous fuels. The nitrogen oxide limits also vary between the different types of solid fuel consumed. This variation in the limits is due to the different amounts of fuel-bound pollutants in each of the different fuel sources.

In 1998, revised nitric oxides emission limits for steam power plants were introduced. In addition to lower limits, the limits for new utility boilers are expressed per MW-hr, which contrasts with the other limits, which are expressed per MMBtu. The use of these new units (MW-hr) is referred to as *output-based* format, where emissions are linked to the amount of power generated. This was used to promote energy efficiency as well as pollution prevention. The use of MMBtu is referred to as an *input-based* format, where emissions are linked to the amount of fuel energy used.

The standards for the stationary gas turbine and Rankine-cycle electric utilities are for that particular source unit and do not cover the ambient air-quality limits. The total pollutant levels for a given region must also be considered. At this point, even facilities not covered by the NSPS are considered, including reciprocating engines. The ambient air-quality is regulated by the national ambient air quality standards (NAAQS) discussed previously, and varies from region to region. The NAAQS determines if a region is governed by the PSD or NA program and, therefore, if BACT or LAER control equipment is required. Different types of pollution control technology are often required for each of three power-generation facilities discussed in this subsection (gas turbine, reciprocating engine, and Rankine cycle), and these are described in numerous references.

As described, the regulations for air pollutants relative to power generation are complex and numerous. To help clarify this situation, Figure 7 is a flow chart of the various regulations and their interactions. As shown, the regulations that govern a given situation depend on several factors. First, the most restrictive limits are applicable. These limitations could be those due to local regulations, or due to original equipment regulations. For local regulations, the national ambient air quality standards (NAAQS) must be met. These limits depend on whether the location is in an attainment zone or a nonattainment zone. Depending on the specific *zone,* the facility might have to meet limits imposed by new source review (NSR), or prevention of significant deterioration (PSD). The technologies

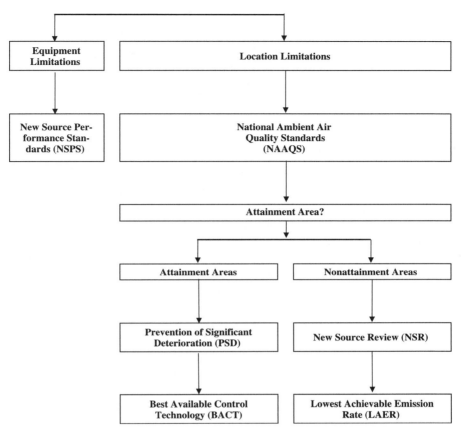

Figure 7 Federal air quality regulations (most restrictive limits apply).

that will be needed for PSD is the best available control technology (BACT); and for the NSR is the lowest achievable emission rate (LAER).

6.5 Water Quality and Solid Waste Disposal

The main legislative basis for managing water pollution is the Federal Water Pollution Control Act of 1956, as amended by the Water Quality Act of 1965, the Federal Water Pollution Control Act amendments of 1972, and the Clean Water Act of 1977. Discharge water will often be monitored, and will often require both state and federal permits if the wastewater is discharged into a public waterway. In some cases, both the temperature and the pH of the discharged water will need to be controlled.

Solid waste disposal is generally not a problem with either natural gas or oil-fired cogeneration plants, but could be a problem for a coal-fired, coal gasification, or waste-to-energy cogeneration system. In some states the bottom ash from waste-to-energy plants has been considered hazardous waste, and the ash disposal cost per ton is more expensive than the refuse disposal cost. All states have different standards for both the quality of the water discharged and the requirements for solid waste disposal, and all project planners should check with the appropriate regulatory agencies in the state where the project is planned to ensure what air, water, and solid waste (if applicable) permits are required.

6.6   Permits and Certificates for Cogeneration

A number of certifications are required to get a cogeneration plant approved. These include not only an FERC certificate but also various state permits. Since each state will set its own permitting requirements, it is not possible to generalize what is required on each application.

As already described, different requirements must be met if the proposed site is in an *attainment* or *nonattainment* zone. An attainment area will require sufficient modeling of ambient air conditions to ensure that no significant deterioration of existing air quality occurs. For a nonattainment area, no new emissions can be added unless they are offset by the removal of existing emissions. This has led to the selling or trading of emissions by industrial facilities and utilities.

Satisfying the emission regulations has become one of the most important factors in determining whether a cogeneration system is feasible for a specific application. The type of prime mover is a major consideration. Gas turbines have often been easier to permit than other prime movers, but each application is unique. Also, the offset of emissions from existing central plant boilers (which are shut down) can often be important in satisfying the overall emission levels.

7   ECONOMIC EVALUATIONS

A decision to install a cogeneration system is often based primarily on economic considerations. Since a cogeneration system requires that funds be spent for the system (capital expense), the monetary earnings and savings resulting from the cogeneration system must be sufficient by some criteria to justify the expenditures. To determine if a cogeneration system is economically justifiable requires a wide range of input information about both the current mode of operation and the proposed system. This section is a brief overview of the considerations that are necessary to complete an economic evaluation of a cogeneration system installation. Fundamentals of completing engineering economic assessments may be found in a number of references (e.g., Ref. 8).

Some of the aspects needed to complete economic evaluations include the computation of charges from the utility for electrical power before and after the installation of the cogeneration system. In addition, both the capital and the operating costs incurred with the installation and operation of a cogeneration system are needed. Finally, all of the economic information must be analyzed to determine the economic feasibility of cogeneration systems. Economic performance measures will be described and compared for use in these economic evaluations.

When discussing and completing economic evaluations for cogeneration systems, several important terms need to be defined. Unlike some engineering projects, installation of a cogeneration system may result in annual savings (e.g., as a result of reducing thermal energy costs) in addition to generating annual income (e.g., as a result of electrical or thermal energy sales). For this reason, economic evaluations for cogeneration systems must be based on revenue, which is defined as the sum of any income and savings cash flows minus the associated costs.

The process of conducting an economic evaluation of a cogeneration system involves nine actions:

1. Determining the current electrical and thermal loads
2. Determining the current or base case operating costs
3. Estimating the future electrical and thermal loads
4. Estimating the future operating costs
5. Estimating the capital costs of the cogeneration system
6. Estimating the new electrical and thermal loads with the cogeneration system
7. Estimating the new operating costs with the cogeneration system
8. Estimating the savings and revenues as a result of the cogeneration system
9. Using this information to complete the overall economic evaluation of the project

For simplicity, the following discussion will consider these items in three categories.

7.1 Operating Costs of Current System

A successful economic analysis must be based on accurate values for current and future (projected) electrical and thermal operating costs. This means that the current and future electrical and thermal loads are needed. Depending on the facility, the appropriate values for the electrical and thermal loads may or may not be easy to obtain. Electrical and thermal loads will generally vary with time and will possess specific profiles on an hourly, daily, monthly, and seasonal basis. For different facilities, the variation may be negligible or significant for one or another of these time periods. On the one hand, in a large industrial plant that operates seven days a week on a 24-hour per day basis, the electrical and process steam loads are often nearly constant. On the other hand, for small manufacturing facilities, schools, and many commercial enterprises, the electrical and thermal loads will vary significantly on hourly, daily, monthly, and seasonal basis. Whenever an accurate economic analysis is required, the electrical and thermal loads need to be obtained for each of these time periods.

For large industrial facilities, the process steam and electric data often are available on an hourly (or even on a 15-minute) basis. For many other facilities, however, obtaining the required load information (particularly on an hourly basis) is difficult. For example, for universities, small manufacturing plants and many commercial enterprises, this type of data is not always available. Typically, whole campus/facility electrical data are available on a monthly basis only (from utility bills), and hourly electrical profiles may have to be constructed from monthly energy and demand data. Hourly thermal data, also, may not exist. Boiler operators may have daily logs, but these may not provide enough detail and may not be in electronic form. Constructing accurate hourly thermal energy profiles can involve hours of tedious work pouring over graphs and boiler operator logs. In the worst-case scenario, only monthly gas bills may be available, and it may be necessary to construct hourly thermal profiles from monthly bills, boiler efficiencies, and heating value content of the fuel. Since the monthly utility charges have to be matched, there is often a great deal of trial and error involved before there is a match between assumed hourly energy profiles and monthly energy consumption and costs from the utility bills.

While first-cut energy and cost analyses may be completed using monthly load profiles, more exact analyses require more detailed load profiles. These more detailed load profiles often consist of hourly data for a typical work day and a typical non-work day, although for an industrial facility that operates seven days a week, 24 hours a day, such detail data may not be needed. That is not the case for many other facilities such as a university, a commercial building complex, or a one- or two-shift manufacturing facility. In these latter cases, the energy loads are highly variable, and hourly analyses are required. In these cases, there will often be a significant difference between weekday and weekend loads. This means that the analysis needs to distinguish between the typical week day and weekend day.

Once a valid energy profile (hourly preferred) is constructed, the current operating costs are determined. This is the annual cost of doing business without the cogeneration system, including annual purchased electrical energy sales, electrical demand charges, boiler fuel costs, boiler maintenance, direct and contract labor, insurance, and other such items. This total number is the baseline to which the cogeneration economics are compared.

Since the economic analysis will be completed for the life of the project, the future electrical and thermal loads are needed. Although these factors are often uncertain, the best estimates should be used. The future electrical and thermal loads should be obtained for each of the years of the cogeneration system life. Based on the future electrical and thermal loads, the projected (non-cogeneration) operating costs should be estimated.

7.2 Operating Costs of the Proposed Cogeneration System

Determining the new electrical and thermal loads associated with a cogeneration system will depend on the type and size of system selected. Ideally, the cogeneration system could be sized to match the electrical and thermal loads exactly; however, seldom is there an exact match.

Based on the new electrical and thermal loads for the cogeneration installation, the projected operating costs should be estimated. For the cogeneration system, these operating costs include the costs of fuels and maintenance. Also, estimates of the new utility costs are needed. Costs of both the electrical energy (per kW-hr) and the electrical demand (kW) are needed. Since utility rates are often dependent on the total power (demand) used, the utility rates and costs may be different, not only because of the reduction of utility electrical energy consumption, but also because of the reduction in electrical demand. Another important cost may be a back-up or stand-by charge imposed by the utility to have power available for the facility in case of an emergency shut-down of the cogeneration system.

As for the current system, the estimated, projected future electrical and thermal loads need to be used for the cogeneration system for each of the project years. Based on these future electrical and thermal loads, the projected operating costs for the cogeneration system should be estimated.

7.3 Economic Merit

By comparing the operating costs of the current system with the costs associated with the cogeneration system, the project savings can be estimated. The savings and income from the cogeneration system must be determined, as well as the additional costs associated with such items as the additional fuel and the additional maintenance. At this point, all the necessary information is available to complete a detailed economic assessment of the cogeneration project for the economic life of the project. Several economic measures may be used to judge the economic feasibility of a specific project.

For some situations, a simple payback (or investor's rate of return) approach is often sufficient. Once an acceptable payback period is defined, a decision can be made on the feasibility of the cogeneration project. If the simple payback period is fairly short (e.g., two to three years), then small variations in the assumptions, approximations, or prices will have little effect on the decision. As the payback period lengthens (e.g., more than four years), then other more complete economic measures should be examined. For these longer periods, the time value of money has to be considered, as well as projected energy rates and projected changes in energy needs for the facility. For projects with these longer payback periods, other economic measures should be used.

An example of a more detailed evaluation would be the use of a net present value analysis. For this approach, all costs and income (including savings) streams are brought to the present using a discount factor and summed. This process starts with the initial investment, which is often the only entry for first year. For each subsequent year, the operating income or savings, depreciation, and taxes are listed. These amounts are used to determine the net cash flow for each year. These values are then adjusted with the discount factor for each year to obtain present values. A final net present value at the end of the project can then be determined and compared to the net present values of other alternatives.

Table 2 is a simple template of the type of final evaluation that may be developed. As shown, tax considerations are a part of the evaluation for for-profit facilities. The number of years will be dictated by the economic life of the project. Typically, the numbers that are needed for each item in Table 3 are supplied from earlier tables that can be arranged to provide the numbers automatically. If setup in such a fashion, a number of studies can be completed rather easily.

Once the first case is completed, various sizing and operating modes for the cogeneration system should then be evaluated. The case with the highest net present value would then be the most favorable economic choice. With any potential operating mode considered, the use of the thermal energy is what generally makes a cogeneration system feasible. If electricity production were the primary output, the cogenerator could not typically compete with the local electric utility. The simultaneous production of and need for the thermal energy is what makes a cogeneration project economically attractive.

A final aspect of any comprehensive economic evaluation is to conduct a series of *sensitivity analyses*. These sensitivity analyses are designed to detect the sensitivity of the results to the assumptions and approximations used in the analysis. For example, the future cost of fuels and electric rates should be varied (by say, ±20 percent) to determine the effect on the final net present value or other economic measures. In a similar manner, the effect of inflation on specific items on the final results should be explored. A comprehensive sensitivity study

Table 2   Template for Final Economic Evaluation

		YEAR 0	YEAR 1	YEAR 2	—
	Investment		—	—	—
Income Calculations	Operating Income	—			
	Depreciation (%)	—			
	Depreciation ($)	—			
	Adjusted Income	—			
Tax Calculations	Tax (%)	—			
	Tax ($)	—			
	Investment Tax Credits (if any)	—			
	Adjusted Tax	—			
Present Value Calculations	Income after Tax (Operating income minus the adjusted tax)	—			
	Discount Factor	—			
	Discounted Revenue	—			
	Net Present Value	—			

Table 3   Summary of the Main Characteristics of the Major Ownership/Financing Structures

Ownership: Operation: Characteristic	Self-owned Self-operated	Not Owned Self-operated	Not Owned Not Self-operated
Typical Financing	Conventional	Leased	Third Party
Capital Requirements	Maximum	None	None
Balance Sheet Impact	Maximum	Some	None
Risk	Maximum	Some	Least
Rate of Return	Depends, probably highest	Depends, probably moderate	Depends, probably lowest
Personnel Required	Maximum	Some	Least
Control of Supply of Electric Power and Thermal	Yes	Yes	No
Fuel Contracts	Required	Required	For others

will include most of the parameters of the economic evaluation, and can be quite extensive.

In general, an economic analysis of a cogeneration system is complicated, and the results will vary greatly depending on values for most of the parameters (such as interest rates, cost of fuel, permitting requirements, cost of electricity sales, and other factors). To complete detailed studies with all of these considerations in a reasonable manner, requires the use of computer programs of one nature or another. Commercial computer programs and simulations for cogeneration systems are available, and spread-sheet type programs can be constructed for specific uses (e.g., Ref. 4).

8   OWNERSHIP AND FINANCIAL ARRANGEMENTS

8.1   Overall Considerations

Even with a reliable technical design and favorable economics, the successful completion of a cogeneration project often will depend on acceptable financial arrangements. Financing is critical to the success of a cogeneration project, and it is best to determine, early on, the financial arrangement to be used. The selected financial arrangement will be intimately linked to the ownership structure. A cogeneration system may be owned by the facility, by a third-party entity, or by a partnership.* The ownership may be structured in a variety of ways. By increasing the number of participants in the project, the individual risk decreases, but the venture is more difficult to organize and the individual potential gains decrease.

---

*In this section, *facility* refers to the entity with the electric and thermal loads. This is sometimes called the thermal consumer, heat consumer, or thermal host. A *third party* refers to an entity separate from the thermal load owner (first party) and from the local utility (second party).

The thermal and electric loads of a facility are the items that might motivate others to be involved in developing a cogeneration system. In one respect, these loads represent the opportunity for financial gains. There are a number of potential project participants that could conceivably have an interest in the development of the cogeneration system. These potential participants include equipment manufacturers, power-plant operators, investors (such as banks, insurance companies, and pension funds), electric utilities, fuel suppliers, engineering firms, and governmental agencies. In general, the goal of the selected ownership structure and the related financial arrangements is to result in some combination of maximum profits, minimal risk, and maximum tax benefits.

External participants, partners, and investors will examine a number of issues before deciding to be involved in the project. These issues include the overall economics, the revenue from thermal and electricity sales, the accuracy of the capital costs and operating expenses, the experience of the participants, the projected availability of the plant, the previous success of the proposed technology, the assurance that permits, contracts, and agreements will be obtained on a timely basis, and the availability and cost of fuel. A net positive assessment of these issues will be necessary for any external participation.

Cogeneration projects can be financed by a variety of options. The traditional approach to financing is owner financing; however, cogeneration facilities are expensive and complex, and a number of alternatives are available for financing. When deciding on the most favorable financial arrangement, the owners of a facility will often consider the following questions:

- Do the owners have adequate capital to finance the whole project?
- If borrowing the money to finance the plant, will the effect on the owners' credit rating be acceptable?
- How will the financing affect the owners' balance sheet?
- Do the owners desire to receive guaranteed savings (and minimize their risk)?
- If the owners finance the project, do they have the ability to utilize available tax benefits?
- Do the owners have an interest in operating or owning the plant (since this is probably not their main line business)?

Answers to these questions will help the owners of a facility select the most attractive ownership and financial structures for their needs. They will evaluate the possible options and select the ones that will have the most favorable impact on their business.

The following are descriptions of examples of possible financial arrangements. They are grouped according to three major ownership structures. For the first category, the owners of the thermal load may own and operate the cogeneration system, and use conventional financing. In the second category, the owners of the thermal load may develop a partnership arrangement for the ownership and financing of the cogeneration system. For the third category, the owners may offer a third-party the opportunity to develop the cogeneration project. A third-party ownership structure has the greatest variety of financing arrangements. Examples of these various financial structures are described next. Other financing arrangements may be possible, including combinations of these, but most financial arrangements will possess characteristics represented by one of more of the following arrangements.

8.2 Conventional Ownership and Operation (100 Percent Ownership)

The owner of the electric and thermal loads has two basic financing options in a conventional owner/operate structure: (1) fund the project internally from profits in other areas of the business, or (2) fund part of the project from internal sources and borrow the remainder from a conventional lending institution. Sole ownership offers the largest degree of control and rewards, but also results in the largest exposure to risk. All external participants (e.g., utilities, engineering and construction firms, and operating and maintenance organizations) must interact with the facility owner.

Most businesses have a minimum internal rate of return on equity that they require for any investment. They may not be willing to fund any project that does not meet the internal hurdle rate using 100 percent equity (internal) financing. With 100 percent internal financing, the company avoids the problems of arranging external financing (perhaps having to add partners). If there is a marginal return of equity, however, the company will not finance the project, especially if the money could be used to expand a product line or create a new product that could provide a greater return on equity.

Because the cost of borrowed money is typically lower than a business's own return on equity requirements, the combination of partial funding internally and conventional borrowing is often used. By borrowing most of the funds, the organization can leverage internal funds for other projects, thus magnifying the overall return on equity.

External contract issues are simpler in conventional ownership. Contracts will be required for the gas supply, for excess power sales to the local utility, possible operating and maintenance agreements, and any agreements with possible lenders. For conventional ownership and financing, no contracts may be needed for the thermal energy and electricity if all is used internally.

8.3 Partnership Arrangements

One alternative to 100 percent ownership is to share ownership with partners. A variety of partnership arrangements are possible. These arrangements include conventional partnerships, limited partnerships, jointly owned corporation, unincorporated association, and others. Partners might include a gas utility, a major equipment vendor (such as a gas turbine manufacturer), investors, and engineering firms.

The major advantage of a partnership is the sharing of risks and credit. The disadvantages are that profits are also shared and that contract complexity increases. Since the partnership is the owner of the cogeneration system, thermal and power sales agreements have to be arranged with the owners of the facility, in addition to the other contracts required under conventional ownership. The joint venture company must develop agreements and contracts with the utilities, engineering and construction firms, possible operating and maintenance firms, and lenders (if any).

8.4 Third-party Ownership

As already mentioned, the potential for financial gain may attract third parties to develop and operate a cogeneration system where the owner of the thermal load would contract for at least a portion of the electrical and thermal outputs of the plant. In third-party ownership, the entity with the electric and thermal loads distances himself from both the financing and construction of the cogeneration facility. A third party arranges the finances, develops the project, arranges for gas supply, sells for any excess power produced, arranges thermal sales to the heat consumer, and contracts operating and maintenance agreements. Under the 1992 National Energy Policy Act, the third party may also be able to enter into an electric power sales contract with the heat consumer as well. The third party might operate the facility and sell the thermal and electric to customers, or the third party might lease the facility to the entity with the electric and thermal loads. Often, the facility is co-located on or in close proximity to the thermal owners' facility.

A number of third-party ownership and financing arrangements are possible. Three of the more common forms are lease arrangement, guaranteed savings arrangement, and energy services contract arrangement.

8.5 Final Comments on Financial Aspects

Financing arrangements are a crucial aspect of most cogeneration developments. These arrangements may range from simple to highly complex. They are affected by internal factors such as ownership arrangements, credit ratings, and risk tolerance. In addition, these financial aspects are affected by external factors such as the financial and credit markets, tax laws, and cogeneration regulations. A variety of initial financial arrangements have been outlined in this section to illustrate the nature of these arrangements. Much more detailed arrangements are possible and often necessary, but these are beyond the scope of this chapter. Table 3 is a summary of the main characteristics of the major ownership/financing structures.

To reduce the costs, risk, and uncertainties associated with financing, a number of actions should be considered. These include establishing reliable and robust contracts and agreements with engineering and construction firms, fuel suppliers, and the local utility. Also, actions to minimize any volatility (or perceived volatility) will enhance the situation and should include using reliable and well-known firms. Other actions that will be useful include emphasizing profitability, flexibility, detail work, and careful understanding of the governing regulations.

9 SUMMARY AND CONCLUSIONS

This chapter has provided a brief overview of cogeneration systems. Cogeneration systems are attractive options for facilities where electrical power and thermal energy are used. The major motivations for considering cogeneration systems are the potential savings in money and energy, and the potential for lower emissions.

The technology for cogeneration exists for a range of sizes, and the procedures to integrate these components into cogeneration systems are well established. The key item in the design of a cogeneration system is the prime mover, which is typically a steam turbine, gas turbine, or reciprocating engine. Other important components are the heat-recovery steam generator (HRSG), possibly adsorption chillers, and other power-plant equipment. The arrangements of the equipment are quite varied, particularly for larger systems. An important aspect of the technical design of a cogeneration system is the selection of the size of the system to match the electrical and thermal energy needs.

In addition to the technical considerations, the application of cogeneration systems involves completing an economic evaluation, and an understanding of the governmental regulations and legislation on electrical power production and on environmental impacts. With respect to electrical power production, certain governmental regulations (PURPA) were passed during the late 1970s that removed barriers and provided incentives to encourage cogeneration development. Finally, no cogeneration assessment would be complete without an understanding of the financial arrangements that are possible, and an understanding of the contracts and agreements that are needed.

REFERENCES

1. M. P. Boyce, "Handbook for Cogeneration and Combined Cycle Power Plants," *American Society of Mechanical Engineers*, New York, 2002.
2. B. F. Kolanowski, "Small-Scale Cogeneration Handbook," Fairmont Press, Lilburn, GA, 2000.
3. S. A. Spiewak and L. Weiss, "Cogeneration and Small Power Production Manual," 5th ed., Fairmont Press, Lilburn, GA, 1999.
4. G. R. Baxter and J. A. Caton "Technical and Economic Assessments of Cogeneration Systems: An Overview of the Development and Application of a Generalized Computer Simulation," Proceedings of the 1997 Power-Gen International Conference, Dallas Convention Center, Dallas, Texas, December 9–11, 1997.

5. R. G. Tessmer, J. R. Boyle, J. H. Fish, and W. A. Martin, "Cogeneration and Wheeling of Electric Power," PennWell Books, Tulsa, OK, 1995.

6. J. A. Caton, N. Muraya, and W. D. Turner "Engineering and Economic Evaluations of a Cogeneration System for the Austin State Hospital," Proceedings of the 26th Intersociety Energy Conversion Engineering Conference, Boston, MA, Paper No. 910444, 5 (August 1991), pp. 438–443.

7. J. H. Horlock, "Cogeneration: Combined Heat and Power—Thermodynamics and Economics," Pergamon Press, Oxford, England, 1987.

8. L. Blank, and A. Tarquin, "Engineering Economy," McGraw-Hill, New York, 2002.

CHAPTER **7**

HYDROGEN ENERGY

E. K. Stefanakos, D. Y. Goswami, S. S. Srinivasan, and J. T. Wolan
Clean Energy Research Center
University of South Florida
Tampa, Florida

1	**INTRODUCTION**	**165**
2	**HYDROGEN PRODUCTION**	**166**
	2.1 Steam Reforming of Natural Gas	166
	2.2 Partial Oxidation of Heavy Hydrocarbons	168
	2.3 Coal Gasification	169
	2.4 Hydrogen Production from Biomass	169
	2.5 Electrolysis	171
	2.6 Thermochemical Processes	173
	2.7 Photoelectrochemical Hydrogen Production	174

	2.8 Biological Methods	176
3	**HYDROGEN STORAGE**	**176**
	3.1 Hydrogen Storage Options	179
4	**HYDROGEN UTILIZATION**	**191**
	4.1 Fuel Cells	191
	4.2 Internal Combustion Engines	196
	4.3 Hydrogen Burner Turbines	197
5	**HYDROGEN SAFETY**	**197**
	5.1 The Nature of Hydrogen	197
	5.2 How to Handle Hydrogen	198
6	**CONCLUSIONS**	**199**

1 INTRODUCTION

Fossil fuels are not renewable. They are limited in supply, their economic cost is continuously increasing, and their use is growing exponentially. Moreover, combustion of fossil fuels is causing global climate change and harming the environment in other ways as well, which points to the urgency of developing environmentally clean alternatives to fossil fuels.

Hydrogen is a good alternative to fossil fuels for the production, distribution, and storage of energy. Automobiles can run on hydrogen either used as fuel in internal combustion engines or fuel-cell cars or in hybrid configurations. Hydrogen is not an energy source but an energy carrier that holds tremendous potential to use renewable and clean energy options. It is not available in free form and must be dissociated from other molecules containing hydrogen, such as natural gas or water. Once produced in free form, it must be stored in a compressed or liquefied form, or in solid-state materials.

There are a number of advantages in using hydrogen as a universal energy medium. The conversion of hydrogen by combustion or fuel cells results in only heat or electricity and water, as represented by the following equation:

$$H_2 + \tfrac{1}{2}O_2 \rightarrow H_2O + \text{Electricity} + \text{Heat} \tag{1}$$

Hydrogen is not toxic and is easily absorbed in the biosphere. It can be readily produced (albeit at high cost) from water by electrolysis. However, additional research is needed for the production from non-fossil resources, storage, and transportation of hydrogen before it becomes commercially viable as an alternative to conventional fuels.

The following sections cover the production, storage, utilization, and safety of hydrogen.

2 HYDROGEN PRODUCTION

As hydrogen is not readily available in its natural state, it must be produced at low cost, without creating any imbalance in global ecology. The conventional technologies used by industry to produce hydrogen are steam reforming of natural gas, partial oxidation of heavy hydrocarbons, gasification, and water electrolysis. All these processes are heavily dependent on fossil fuels. Thus, they have the inherent pollution and availability problems. The other potential ways of producing hydrogen include photoelectrochemical, photochemical, thermochemical, and biological methods using renewable energy sources.

2.1 Steam Reforming of Natural Gas

Steam reforming of natural gas, or *steam methane reformation* (SMR), is one of the most developed and commercially used technologies. A block diagram of the SMR process is shown in Figure 1. Steam reforming of natural gas involves two steps. The first step is for the feedstock consisting of light hydrocarbons, usually methane, to react with steam at elevated temperatures (700°C–925°C) to produce *syngas*—a mixture of hydrogen (H_2) and carbon monoxide (CO). The process is endothermic, and heat of reaction is supplied by the combustion of fossil fuels. This process requires a catalyst inside the reformer for the reactions to occur. To protect the catalyst from corrosion, the feedstock must pass through a desulfurization process prior to entering the reformer. The second step, a water–gas shift reaction, reacts carbon monoxide with steam to produce additional H_2 and carbon dioxide (CO_2) at around 350°C. This reaction is known as a shift reaction and is used to increase the H_2 content. Finally, a mixture of CO_2 and H_2 is sent to a gas purifier, where the hydrogen is separated from CO_2 via one of many methods (pressure swing absorption, wet scrubbing, or membrane separation). The chemical reactions involved in the SMR process are shown in

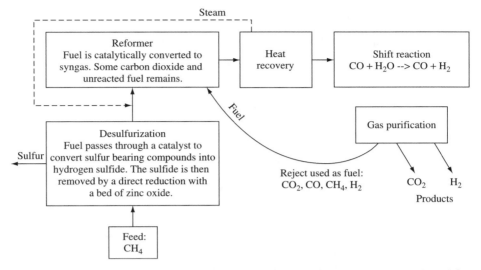

Figure 1 Block diagram of hydrogen production by steam reforming process. (Adapted from Ref. 3.)

equations (2) and (3):

$$C_nH_m + nH_2O \rightarrow nCO + \left(\frac{2n + m}{2}\right)H_2 \qquad (2)$$

$$CO + H_2O \rightarrow CO_2 + H_2 \qquad (3)$$

For effective H_2 production using SMR, high temperatures at the reformer exit and an excess of steam to the reactor are required. Temperatures of 800°C to 900°C and a molar steam to carbon ratio of $S/C = 2.5$ to 3.0 are considered the optimum conditions.[1] Overall, SMR produces hydrogen with a purity of 96 to 98 percent[2] and with operating efficiencies ranging from 65 to 75 percent, as estimated by Sherif et al.[3]

SMR is the most widely used and cheapest process for producing H_2 and is used to produce 48 percent of the world's hydrogen.[4] The price of hydrogen production from the SMR process strongly depends on the cost and availability of the natural gas feedstock. Kirk and Ledas estimated that feedstock cost contributes 52 to 68 percent of the overall hydrogen production expense.[5] Basye and Swaminathan reported the cost division as 60 percent on feedstock, 30 percent from capital related charges and 10 percent owing to operation and maintenance costs.[6]

The SMR process is heavily dependent on fossil fuels. Moreover, it is not 100 percent efficient and some of the energy value of the hydrocarbon fuel is lost while in the conversion to hydrogen. The emissions of CO_2 from the SMR process can be reduced to some extent by sequestering the CO_2 (by storing it underground or in canisters). This capture and storage of CO_2 increases the capital and operational costs by about 25 to 30 percent.[7] However, according to Padro and Putsche, even after the inclusion of these costs, SMR is still less expensive than producing hydrogen from electrolysis (using large scale hydro power plants).[8] It has also been found with the SMR process that heavier feedstocks (e.g., oil) cannot be used to supply the reformer, due to the need for the feed to be vapor.[9] The possible areas of efficiency improvement include pre-reformers and medium temperature shift reactors.[9]

2.2 Partial Oxidation of Heavy Hydrocarbons

Partial oxidation (POX) refers to the conversion of heavy hydrocarbon feedstocks (e.g., residual oil from the treatment of crude oil) into a mixture of H_2, CO, and CO_2 using superheated steam and oxygen. Figure 2 provides a schematic representation of the POX process. The external energy required to drive the process is obtained through the combustion of the feedstock itself. This necessitates controlling the quantity of O_2 and water vapor required for the reactions. To increase the H_2 content, the mixture of H_2, CO, and CO_2 is subjected to the shift reaction. This results in the formation of H_2 and CO_2. The reactions involved in a POX process are as following:

$$C_nH_m + \left(\frac{n}{2}\right)O_2 \rightarrow nCO + \left(\frac{m}{2}\right)H_2 + \text{heat} \tag{4}$$

$$C_nH_m + nH_2O + \text{heat} \rightarrow nCO + \left(n + \frac{m}{2}\right)H_2 \tag{5}$$

$$CO + H_2O \rightarrow CO_2 + H_2 + \text{heat} \tag{6}$$

where $n = 1$ and $m = 1.3$ for residual oils.[3] POX produces hydrogen with a purity of 96 to 98 percent.[2]

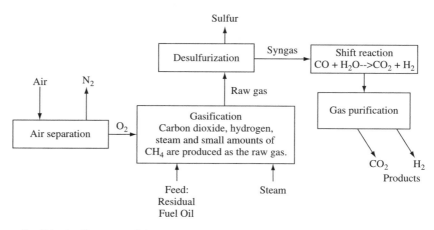

Figure 2 Block diagram of hydrogen production by partial oxidation. (Adapted from Ref. 3.)

The process works with any liquid or gaseous hydrocarbon. The overall efficiency of the process is about 50 percent.[3,8] Like natural gas reforming, production cost of hydrogen is influenced by the price of feedstock. Some problems with POX include the need for an air separation unit. The air separation unit is needed to supply the pure oxygen to the process in order to prevent the release of nitrous oxide to the environment.[6] The addition of this unit increases the system's capital cost thus increasing the hydrogen product cost.

2.3 Coal Gasification

Gasification is similar to partial oxidation except it has two main differences: gasification occurs at much higher temperature ($1,100°–1,300°C$), and it uses a wide range of solid feedstocks (coal, heavy refinery residuals, biomass). In this process, a dry or slurried form of the feedstock is subjected to elevated temperature and pressure conditions in an oxygen-starved environment. This leads to an efficient and clean conversion of carbonaceous substances into a mixture of gas, containing mainly carbon monoxide and hydrogen. Inorganic materials in the feed are finally removed as a molten slag at the bottom of the reactor. The entire gasification process can be represented by the following reactions:

$$2C + O_2 \rightarrow 2CO \tag{7}$$

$$C + H_2O \rightarrow CO + H_2 \tag{8}$$

$$C + CO_2 \rightarrow 2CO \tag{9}$$

Coal is the most abundant fossil fuel. Gasification of coal offers higher thermal efficiencies than conventional coal-fired power generation and also has less impact on the environment. Low-grade coal types can be effectively used in coal gasification, expanding the available fossil-fuel options.

2.4 Hydrogen Production from Biomass

Biomass represents a large potential feedstock resource for environmentally clean hydrogen production. It lends itself to both biological and thermal conversion processes. In the thermal path hydrogen can be produced in two ways: *direct gasification* and *pyrolysis* to produce liquid bio-oil, followed by steam reforming.

Direct gasification of biomass is in many ways similar to coal gasification. The process occurs broadly in three steps:

1. Biomass is gasified (using steam or air) to produce an impure syngas mixture composed of hydrogen, CO, CO_2, CH_4, small amounts of higher hydrocarbons, tar, and water vapor. The gas may also contain particulate matter, which is removed using cyclones and scrubbers. The particulate free gas is compressed and then catalytically steam reformed to eliminate the tars and higher hydrocarbons.

2. High- and low-temperature shift conversions convert the CO to CO_2 and thereby produce additional hydrogen.

3. The hydrogen is separated from other products by PSA (Pressure Swing Adsorption).[10]

Figure 3 illustrates the sequence of processes. The main reactions taking place in biomass gasification are as follows:

$$C_nH_mO_l + H_2O/O_2 \rightarrow H_2 + CO + CO_2 + C_nH_m + tars + C(s) + \Delta H_R$$

$$\Delta H_R > 0 \quad \text{(biomass gasification)} \qquad (10)$$

$$CO + H_2O \rightarrow CO_2 + H_2$$

$$\Delta H_R = -41.2 \text{ kJ/mol} \quad \text{(water-gas shift)} \qquad (11)$$

Biomass typically contains about 6 percent hydrogen by weight. However, in the presence of hydrogen-bearing species (steam), the hydrogen yield can be considerably improved above the 6 percent minimum.[11] Gasification temperatures encountered are typically in the range $600°$ to $850°C$, which is lower than many thermochemical water-splitting cycles thereby making biomass gasification an attractive technology to produce hydrogen. Steam gasification of biomass is endothermic. The energy required for the process is supplied by burning part of the biomass feedstock or uncombusted char. Tars are polyaromatic hydrocarbons produced during gasification of biomass. However, tars are undesirable co-products, as they clog filters, pipes, and valves and damage downstream equipments such as engines and turbines. Efforts are being made to minimize or reform the tars to additionally produce hydrogen.[12,13]

Hydrogen can alternately be produced by reforming the biomass to a liquid bio-oil in a process called *pyrolysis*. Pyrolysis is an endothermic thermal decomposition of biomass carried out in an inert atmosphere at $450°$ to $550°C$.[14] The bio-oil so produced is a liquid composed of 85 percent oxygenated organics and 15 percent water. The bio-oil is then steam-reformed in the presence of a

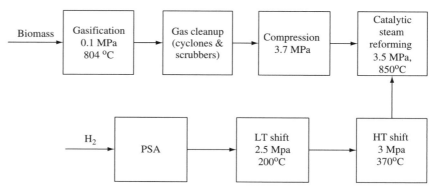

Figure 3 Gasification followed by steam reforming (From Ref. 10.)

nickel-based catalyst at 750° to 850°C, followed by shift conversion to convert CO to CO_2.[15] The reactions can be written as follows:

$$Biomass \rightarrow Bio - Oil + char + gas \quad (pyrolysis) \tag{12}$$

$$Bio = Oil + H_2O \rightarrow CO + H_2 \quad (reforming) \tag{13}$$

$$CO + H_2O \rightarrow CO_2 + H_2 \quad (water\text{-}gas\ shift) \tag{14}$$

2.5 Electrolysis

Hydrogen production by electrolysis of water is a mature and efficient technology. The electrodes are separated by an ion-conducting electrolyte, as shown in Figure 4. Hydrogen and oxygen are produced at the cathode and the anode, respectively. To keep the produced gases isolated from each other, an ion-conducting diaphragm is used to separate the two chambers.

Equation 15 shows the overall chemical equation for electrolysis:

$$2H_2O + Energy\ (Electricity) \rightarrow O_2 + 2H_2 \tag{15}$$

Reaction at the anode

$$H_2O \rightarrow 0.5O_2 + 2H^+ + 2e^- \tag{16}$$

Reaction at the cathode

$$2H^+ + 2e^- \rightarrow H_2 \tag{17}$$

The reversible decomposition potential of equation (15) is 1.229 V at standard conditions of 1 atm pressure and 25°C. However, the total theoretical water decomposition potential is 1.480 V corresponding to hydrogen's enthalpy. The actual potential is typically between 1.75 and 2.05 V due to irreversibility's and internal resistance. Typical efficiencies are of the order of 80 percent.

Electrolysis cells are normally characterized by their electrolytes (e.g. alkaline electrolyzer, solid polymer electrolyte (SPE) electrolyzer, or solid oxide electrolyzer).

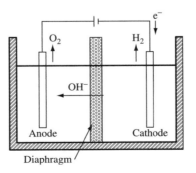

Figure 4 Electrolysis of water in an alkaline electrolyzer.

Alkaline Water Electrolyzer

Alkaline water electrolysis is the most common type of electrolysis currently in use for large-scale electrolytic hydrogen production. The most common electrolyte used in alkaline water electrolysis is aqueous potassium hydroxide (KOH) at 30 percent concentration owing to the high conductivity and high resistance to corrosion of stainless steel at this concentration.[16] These electrolyzers work effectively under the operating conditions of 70° to 100°C and 1 to 30 bar. Asbestos has commonly been used as a diaphragm material to prevent hydrogen and oxygen gases from mixing together inside the cell. The principle of alkaline water electrolysis is shown schematically in Figure 5. Commercial alkaline water electrolyzers are typically classified into two main types, unipolar and bipolar.[17]

Solid Polymer Electrolyte (SPE) Electrolyzer

SPE electrolyzers, as the name suggests, use solid polymer electrolytes, which are made up of special materials also called *perfluorocarbon ion exchange membranes*. This ion exchange membrane is sandwiched between catalyst-loaded electrodes. Water is fed to the anode of an electrolysis cell, which is generally made up of porous titanium and activated by a mixed noble metal oxide catalyst. At the anode, water splits into oxygen and protons. The protons migrate through the ion exchange membrane to the cathode, where they are reduced to hydrogen. Figure 5 shows a simplified schematic of the SPE electrolyzer. SPE electrolyzers are also referred to as proton or polymer exchange membrane (PEM) electrolyzers.

The most common proton-conducting solid electrolytes are perfluoroalkyl sulfonic acid polymers, such as Nafion®. Because of the dehydration of the membrane, the operating temperature of the SPE devices is limited to about 80°C.[18] To raise the dehydration temperature, several aromatic sulfonic acid polymers were synthesized and characterized. These were polyetheretherketone (PEEK), polyethersulfone (PES), polyphenylquinoxaline (PPQ), and polybenzimidazole (PBI).

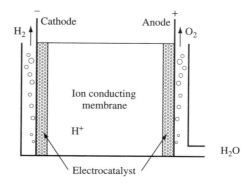

Figure 5 Schematic representation of a SPE Electrolyzer.

2.6 Thermochemical Processes

Thermochemical hydrogen production is a means of splitting water via a series of chemical reactions. All chemical intermediates are recycled internally within the process so that water is the only raw material and hydrogen and oxygen are the only products. The maximum temperature requirements for most thermochemical cycles lie within a temperature range of 650° to 1,100°C, thus eliminating use of lower temperature heat sources.[19] Figure 6 illustrates the concept of splitting water by a thermochemical cycle.

Wendt claims that 2,000 to 3,000 different theoretical cycles have been proposed and evaluated.[16] They can all be subdivided into four basic steps: water-splitting reaction; hydrogen production; oxygen production; and material regeneration.

The ability to reuse almost all of the components involved in the cycle (except feedwater) makes the thermochemical process attractive. Among the 2,000 to 3,000 possible thermochemical cycles, Fewer than ten have been studied extensively. The important ones under research now are described below.

ZnO/Zn Cycle

The ZnO/Zn Cycle is a two-step water-splitting sequence based on the thermal redox pairs of metal oxides. As shown in the following reactions, the process relies on the endothermic thermal dissociation of ZnO, followed by the exothermic hydrolysis of Zn:

$$ZnO \xrightarrow{heat} Zn(g) + 0.5O_2 \tag{18}$$

$$Zn + H_2O \rightarrow ZnO + H_2 \tag{19}$$

Theoretical conversion efficiencies for this cycle are of the order of 50 percent.[20] This cycle is in its preliminary stage of development. The current work is focused on the quenching step and decomposition rate measurements. The

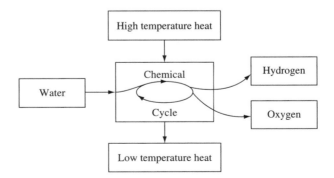

Figure 6 Schematic diagram of thermochemical cycle. (Adapted from Ref. 19.)

success of this cycle will depend on the development of an efficient mechanism for separating the Zn–H_2 mixture.

UT-3 Cycle

The UT-3 Cycle, invented at University of Tokyo, consists of Ca, Fe, and Br compounds. It involves the use of solid and gaseous phases of reactants and products. The predicted first law and second law efficiencies of the adiabatic UT-3 cycle are 49 percent and 53 percent respectively.[21] This cycle is composed of the following reactions:

$$2CaO(s) + 2Br_2(g) \xrightarrow{680^\circ C} 2CaBr_2(s) + O_2(g) \tag{20}$$

$$CaBr_2(s) + H_2O(g) \xrightarrow{760^\circ C} CaO(g) + 2HBr(g) \tag{21}$$

$$3FeBr_2(s) + 4H_2O(g) \xrightarrow{560^\circ C} Fe_3O_4(s) + 6HBr(g) + H_2(g) \tag{22}$$

$$Fe_3O_4(s) + 8HBr(g) \xrightarrow{210^\circ C} 3FeBr_2(s) + 4H_2O(g) + Br_2(g) \tag{23}$$

A substantial amount of research work, including bench scale laboratory tests,[23] solid reactants development, and reaction kinetic measurements have been performed on this cycle.[24,25,26]

Iodine-Sulfur Cycle

The Iodine-Sulfur cycle is a three-step thermochemical process for decomposing water into H_2 and O_2. The cycle proposed by General Atomic Co. is one of the most extensively studied cycles. The chemical reactions involved in the process are shown in equations 23 to 25.[26]

$$xI_2(l) + SO_2(aq) + 2H_2O(l) \xrightarrow{20-100^\circ C} 2HI_x(l) + H_2SO_4(aq) \tag{24}$$

$$2HI(l) \xrightarrow{200-700^\circ C} H_2(g) + I_2(g) \tag{25}$$

$$H_2SO_4(g) \xrightarrow{850^\circ C} H_2O(g) + SO_2(g) + 0.5O_2(g) \tag{26}$$

The maximum hydrogen production efficiency as estimated by GA was around 50 percent.[27] Practical implementation of this cycle has been limited by the effective separation of HI and H_2 from HI–I_2–H_2O and H_2–H_2O–HI–I_2 mixtures, respectively. Moreover, the severe corrosion issue associated with high-temperature hydriodic acid and sulfuric acid needs to be addressed.

2.7 Photoelectrochemical Hydrogen Production

Photoelectrochemical (PEC) systems combine both photovoltaics and electrolysis into a one-step water-splitting process. These systems use a semiconductor

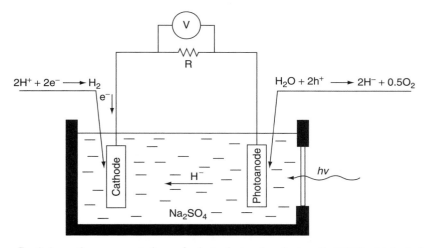

Figure 7 Schematic representation of photoelectrochemical cell (PEC). (Adapted from Ref. 28.)

electrode exposed to sunlight in combination with a metallic or semiconductor electrode to form a PEC cell. A schematic illustration of a PEC hydrogen production driven by solar energy is shown in Figure 7.[28]

In general, the semiconductor electrode (photoanode) is activated by solar radiation, which drives the reaction in an aqueous solution. Looking closely into the reaction involved in the photoelectrochemical process, we find that, due to band gap illumination, electrons and holes are formed in conduction and valence band, respectively, at the photoanode.

$$2h\upsilon \rightarrow 2e^- + 2h^+ \tag{27}$$

where h is the Plank's constant, υ the frequency, e^- the electron, and h^+ the hole.

The photogenerated holes at the anode split the water molecules into hydrogen ions and oxygen. The released hydrogen ions migrate to the cathode through the aqueous electrolyte.

$$2h^+ + 2H_2O \rightarrow \tfrac{1}{2}O_2 + 2H^+ \tag{28}$$

Electrons generated at photoanode are transferred over the external circuit to the cathode, where they reduce hydrogen ions into gaseous hydrogen.

$$2H^+ + 2e^- \rightarrow H_2 \tag{29}$$

Thus, solar energy is utilized to produce hydrogen. Ideally, electrochemical decomposition of water takes place when the electromotive force of the cell is equal to 1.23 eV. If we consider internal losses in the PEC, a minimum band gap of 1.8 eV is required to run the reaction.[29]

The PEC devices have an advantage over conventional PV, since they don't require semiconductor/semiconductor p–n junctions. In PEC, the junction is

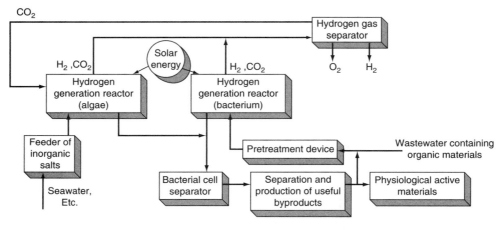

* Adapted from Miyake et al. *Journal of Biotechnology* **70** (1999): p.90, Figure 1.

Figure 8   Concept of biological hydrogen production.

formed intrinsically at the semiconductor/electrolyte interface. Chemical photocorrosion and high costs have so far prevented the commercial utilization of photoelectrochemical devices.

2.8   Biological Methods

Hydrogen can also be obtained from biological process involving organic compounds. Figure 8 shows one of the techniques for hydrogen production using biological means.

There are two fundamental ways of biological hydrogen production:

1. Fermentation of the bacteria, which is an anaerobic process that converts organic substances such as starch, cellobiose, sucrose, and xylose to H_2 and CO_2 without the need of sunlight and oxygen
2. Biophotolysis, a process that uses micro-algae-cynobacteria and green algae to produce hydrogen in the presence of sunlight and water

Both of these processes are being researched.

3   HYDROGEN STORAGE

An intermediate storage of hydrogen is mandatory for on-board vehicular applications, and this is schematically represented in Figure 9.[30,31,32,33,34].

The barriers and limitations of the existing and new hydrogen storage technologies (Table 1) delay its practical use for commercial applications.[35] The development and commercialization of new technologies are required to meet the US-DOE milestones (Table 2) and FreedomCAR technical target performance (Table 3).[36,37]

Table 1 Hydrogen Storage Methods and Phenomena (gravimetric density, ρ_m, the volumetric density ρ_v, the working temperature T, and presure P)

Storage Method	ρ_m (mass %)	ρ_v (kg H$_2$ m^{-3})	T ($^{\circ}$K)	P (bar)	Phenomena and Remarks
High-pressure gas cylinders	13	<40	303	800	Compressed gas (molecular H$_2$) in light weight composite cylinders (tensile strength of the material is 2000 MPa)
Liquid hydrogen in cryogenic tanks	Size dependent	70.8	21	1	Liquid hydrogen (molecular H$_2$) continuous loss of a few % per day of hydrogen at room temperature
Adsorbed hydrogen (carbon nanotube)	2	20	193	100	Physisorption (molecular H$_2$) on materials (e.g., carbon with a very large specific surface area), reversibility problems
Absorbed on interstitial sites in a host metal (Metal hydrides)	2	150	303	1	Hydrogen (atomic H) intercalation in host metals, metallic hydrides working at room temperature are fully reversible
Complex compounds	<18	150	>373	1	Complex compounds ([AlH$_4$]$^-$ or [BH$_4$]$^-$), desorption at elevated temperature, absorption at high pressures
Metals and complexes together with water	<40	>150	303	1	Chemical oxidation of metals with water and liberation of hydrogen, not directly reversible

Table 2 US-DOE Hydrogen Storage Milestones

Targets	2010	2015
System gravimetric capacity = "specific energy"	6 wt.%; 7.2 MJ/kg; 2.0 kWh/kg	9 wt.%; 10.8 MJ/kg; 3.0 kWh/kg
System volumetric capacity = energy density	1.5 kWh/L; 5.4 MJ/L; 45 g/L	2.7 kWh/L; 9.7 MJ/L; 81 g/L
Storage system cost	$4/kWh ($133/kg H$_2$)	$2/kWh; ($67/kg H$_2$)

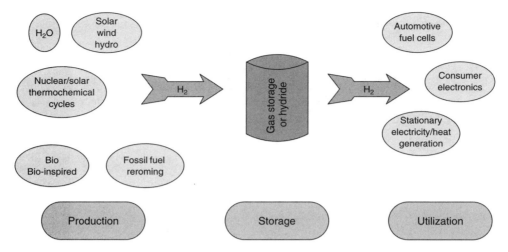

Figure 9   The hydrogen economy consists of production, storage, and use.

Table 3   FreedomCAR Hydrogen Storage System Targets

Targeted Factor	2005	2010	2015
Specific energy (MJ/kg)	5.4	7.2	10.8
Hydrogen (wt.%)	4.5	6.0	9.0
Energy density (MJ/L)	4.3	5.4	9.72
System cost ($/kg/system)	9	6	3
Operating temperature (°C)	−20/50	−20/50	−20/50
Cycle life-time (absorption/desorption cycles)	500	1,000	1,500
Flow rate (g/s)	3	4	5
Delivery pressure (bar)	2.5	2.5	2.5
Transient response (s)	0.5	0.5	0.5
Refueling rate (kg H_2/min)	0.5	1.5	2.0

Several critical properties of the hydrogen storage materials can be evaluated for automotive applications:

- Light weight
- Cost and availability
- High volumetric and gravimetric density of hydrogen
- Fast kinetics
- Ease of activation
- Low temperature of dissociation or decomposition

- Appropriate thermodynamic properties
- Long-term cycling stability
- High degree of reversibility

3.1 Hydrogen Storage Options

Current hydrogen storage technologies include high-pressure tanks, cryogenic storage, metal hydrides, chemical hydrides, and high surface adsorbents such as nanostructured carbon-based materials. High pressure and cryogenic tanks, high surface adsorbents, and many metal hydrides fall in the category of reversible on board hydrogen storage, since refueling with hydrogen can take place directly on board the vehicle. For chemical hydrogen storage and some high-temperature metal hydrides, hydrogen regeneration is not possible on board the vehicle, and thus these systems must be regenerated off board (see Figure 10).

Hydrogen can be stored as a gas or liquid in pressure vessels. Gaseous storage requires large volume and pressure (up to 10,000 psi). Liquid storage requires low temperatures (-423°C) with cryogenic systems. Hydrogen can also be stored in advanced solid state materials—within the structure or on the surface of certain materials, as well as in the form of chemical precursors that undergo a chemical reaction to release hydrogen.[38,39,40,41] Figures 11 and 12 demonstrate these processes in the atomic or molecular scale.

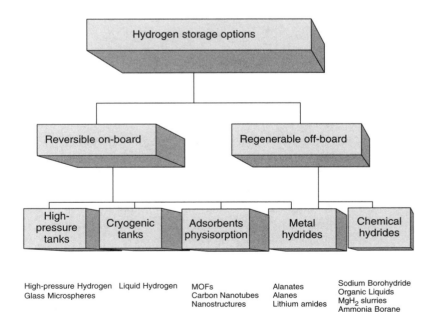

Figure 10 Options for vehicular hydrogen storage.

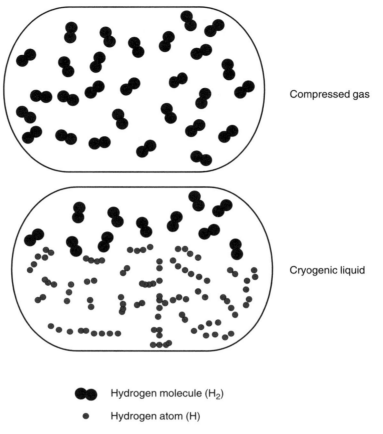

Figure 11 Different types of hydrogen storage methods.

The volume storage efficiencies of gaseous and liquid hydrogen storage are generally very low compared to the solid state hydrogen storage.[42] Figure 12 shows the hydrogen storage methods in solids (by adsorption) or within solids (by absorption). In *adsorption*, (a), hydrogen attaches to the surface of a material either as hydrogen molecules (H_2) or hydrogen atoms (H). In *absorption*, (b), hydrogen molecules dissociate into hydrogen atoms that are incorporated into the solid lattice framework. This method may make it possible to store large quantities of hydrogen in smaller volumes at low pressure and room temperature. Finally, hydrogen can be bound strongly within molecular structures as chemical compounds containing hydrogen atoms, (c).

High-pressure Gaseous Hydrogen Storage
The energy density of gaseous hydrogen can be improved by storing hydrogen at higher pressures. This requires material and design improvements in order to

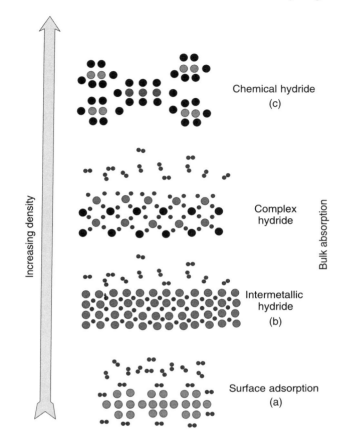

Figure 12 Hydrogen storage in carbon, metal/complex hydrides, and chemical compounds.

ensure tank integrity. Advances in compression technologies are also required to improve efficiencies and reduce the cost of producing high-pressure hydrogen.

Carbon fiber–reinforced 5,000-psi and 10,000-psi compressed hydrogen gas tanks (Figure 13) are under development by Quantum Technologies and others.[43] Such tanks are already used in prototype hydrogen-powered vehicles.

The inner liner of the tank is a high-molecular-weight polymer that serves as a hydrogen-gas permeation barrier. A carbon fiber-epoxy resin composite shell is placed over the liner and constitutes the gas pressure load-bearing component of the tank. Finally, an outer shell is placed on the tank for impact and damage resistance. The pressure regulator for the 10,000 psi tank is located in the interior of the tank. There is also an in-tank gas sensor to monitor the tank temperature during the gas-filling process when the tank is heated.

Issues with compressed hydrogen gas tanks revolve around high pressure, weight, volume, conformability, and cost. The cost of high-pressure compressed gas tanks is essentially dictated by the cost of the carbon fiber that must be

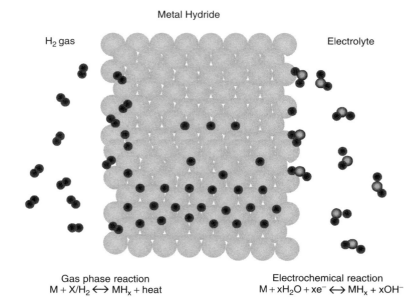

Figure 13   Hydrogen in atomic form on interstitial lattice sites of an intermetallic alloy.

used for lightweight structural reinforcement. Efforts are underway to identify lower-cost carbon fibers that can meet the required high pressure and safety specifications for hydrogen gas tanks. However, lower-cost carbon fibers must still be capable of meeting tank thickness constraints in order to help meet volumetric capacity targets. Thus, lowering cost without compromising weight and volume is a key challenge.

Two approaches are being pursued to increase the gravimetric and volumetric storage capacities of compressed gas tanks from their current levels. The first approach involves cryo-compressed tanks. This is based on the fact that, at fixed pressure and volume, gas tank volumetric capacity increases as the tank temperature decreases. Thus, cooling a tank from room temperature to liquid nitrogen temperature ($77°$K) will increase its volumetric capacity by a factor of four, although system volumetric capacity will be less than this due to the increased volume required for the cooling system.

The second approach involves the development of conformable tanks. Current liquid gasoline tanks in vehicles are highly conformable in order to take maximum advantage of available vehicle space. Concepts for conformable tank structures are based on the location of structural supporting walls. Internal cellular-type load-bearing structures may also be a possibility for greater degree of conformability.

Compressed hydrogen tanks [5,000 psi ($\sim$35 MPa) and 10,000 psi ($\sim$70 MPa)] have been certified worldwide according to ISO 11439 (Europe), NGV-2 (U.S.), and Reijikijun Betten (Iceland) standards and approved by TUV (Germany) and the High-Pressure Gas Safety Institute of Japan (KHK). Tanks have been demonstrated in several prototype fuel cell vehicles and are commercially available. Composite, 10,000 psi tanks have demonstrated a 2.35 safety factor (23,500 psi burst pressure), as required by the European Integrated Hydrogen Project specifications.

Liquid Hydrogen Storage

The energy density of hydrogen can be improved by storing hydrogen in a liquid state. However, the issues with LH_2 tanks are hydrogen boil-off, the energy required for hydrogen liquefaction, volume, weight, and tank cost.[44,45] There are four contributing mechanisms to boil-off losses in cryogenic hydrogen storage systems:

1. Ortho-para conversion
2. Heat leak (shape and size effect, thermal stratification, thermal overfill, insulation, conduction, radiation, cool-down)
3. Sloshing
4. Flashing

Typically, 30 percent of the heating value of hydrogen is required for liquefaction. New approaches, which can lower these energy requirements and thus the cost of liquefaction, are needed. Hydrogen boil-off must be minimized or eliminated for cost, efficiency, and vehicle range considerations, as well as for safety considerations when vehicles are parked in confined spaces. Insulation is required for LH_2, tanks, and this reduces system gravimetric and volumetric capacity.

Liquid hydrogen (LH_2) tanks can store more hydrogen in a given volume than compressed gas tanks. The volumetric capacity of liquid hydrogen is 0.070 kg/L, compared to 0.030 kg/L for 10,000 psi gas tanks. Liquid tanks are being demonstrated in hydrogen-powered vehicles, and a hybrid tank concept combining both high-pressure gaseous and cryogenic storage is being studied.

Metal/Complex Hydrides

Hydrogen can be packed and stored in a solid state by forming a metal hydride.[44–52] During the formation of the metal hydride, hydrogen molecules are dissociated into hydrogen atoms, which insert themselves into interstitial spaces inside the lattice of intermetallic compounds and/or alloys (Figure 13). The typical reversible metal-hydrogen interaction occurs either as a gas-phase reaction or as an electrochemical reaction.

In such a way, an effective storage comparable to the density of liquid hydrogen is created. However, when the mass of the metal or alloy is taken into account, the metal hydride gravimetric storage density is comparable to storage of pressurized hydrogen. The best achievable gravimetric storage density is

Table 4 Theoretical Capacities of Hydriding Substances as Hydrogen Storage Media

Medium	Hydrogen Content kg/kg	Hydrogen Storage Capacity, kg/liter of vol.	Energy Density kJ/kg	Energy Density kJ/liter of vol.
MgH_2	0.070	0.101	9,933	14,330
Mg_2NiH_4	0.0316	0.081	4,484	11,494
VH_2	0.0207		3,831	
$FeTiH_{1.95}$	0.0175	0.096	2,483	13,620
$TiFe_{0.7}Mn_{0.2}H_{1.9}$	0.0172	0.090	2,440	12,770
$LaNi_5H_{7.0}$	0.0137	0.089	1,944	12,630
$R.E.Ni_5H_{6.5}$	0.0135	0.090	1,915	12,770
Liquid H_2	1.00	0.071	141,900	10,075
Gaseous H_2 (100 bar)	1.00	0.0083	141,900	1,170
Gaseous H_2 (200 bar)	1.00	0.0166	141,900	2,340
Gasoline	−	−	47,300	35,500

about 0.07 kg of H_2/kg of metal, for a high-temperature hydride such as MgH_2, as shown in Table 4, which gives a comparison of some hydriding substances with liquid hydrogen, gaseous hydrogen, and gasoline.[10]

The potential to use hydrides for energy storage and applications has stimulated extensive theoretical and experimental research on the fundamental aspects of hydrogen sorption, and on several reversible storage intermetallics such as FeTi,[53] LaNi_5, MmNi_{4.5}Al_{0.5},[54] and Mg_2Ni.[55] Since the maximum weight percentage storage for these intermetallics is ~1.8 wt.% at ambient conditions and ~3.8 wt.% at high temperature (300° to 400°C), there is on-going research to find better hydride materials with higher storage capacity at ambient as well as high-temperature conditions. To achieve this, two prominent routes are being followed: first, to modify and optimize the current storage materials such as FeTi, LaNi_5, and the high-temperature hydride Mg_2Ni, and second, to develop altogether new storage materials, such as transition metal complexes, composite materials, nano-particle and nano-structured materials, and new carbon variants (fullerenes, C_{60} and other higher versions, graphitic nanofibers, and nanotubes).

Magnesium (Mg) has the highest theoretical hydrogen storage capacity of ~7.6 wt.%. However, it has two significant disadvantages: (1) the Mg-H_2 reaction has poor kinetics; and, (2) the resulting hydride is not reversible under ambient or moderate temperature and pressure conditions.[56] A possible way to achieve Mg like storage capacity but with reversible hydrogenation characteristics is to form composites with Mg as one of the components. The other component may be one of the known hydrogen storage intermetallic alloys.[57]

An important feature of the metallic hydrides is the high volumetric density of the hydrogen atoms present in the host lattice as shown in Figure 14.[58]

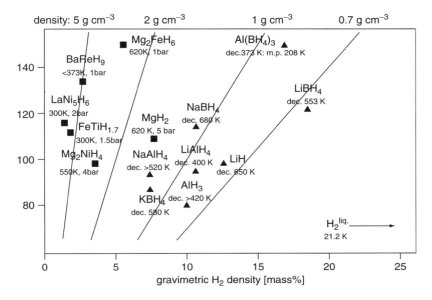

Figure 14 Volumetric and gravimetric hydrogen density of some selected hydrides. Mg_2FeH_6 shows the highest known volumetric hydrogen density of 150 kg H_2 m^{-3}.

The highest theoretical volumetric hydrogen density known today is $150\,kg\,m^{-3}$ for Mg_2FeH_6 and $Al(BH_4)_3$. The Mg_2FeH_6 hydride belongs to the family of Mg-transition metal complex hydrides with $[FeH_6]^{4-}$ octahedral surrounded by Mg atoms in cubic configuration.[59] Interestingly, iron does not form intermetallic compounds with Mg, but it readily combines with hydrogen and Mg to form ternary hydride Mg_2FeH_6 according to these reactions:

$$2Mg + Fe + 3H_2 \rightleftarrows Mg_2FeH_6 \tag{30}$$

$$2MgH_2 + Fe + H_2 \rightleftarrows Mg_2FeH_6 \tag{31}$$

During the storage process (charging or absorption) heat is released, which must be removed in order to achieve the continuity of the reaction. During the hydrogen release process (discharging or desorption), heat must be supplied to the storage tank. The thermodynamic aspects of hydride formation from gaseous hydrogen are described by means of pressure-composition isotherms, as shown in Figure 15. While the solid solution and hydride phase coexist, the isotherms show a flat plateau, the length of which determines the amount of H_2 stored. The stability of metal hydrides is usually presented in the form of Van't Hoff plots (ln P vs. T^{-1}). The most stable binary hydrides have enthalpies of formation of $\Delta H_f = -226\,kJ\,mol^{-1}\,H_2$, the least stable hydrides having enthalpies of formation of $+ 20\,kJ\,mol^{-1}\,H_2$.[60] An advantage of storing hydrogen in hydriding substances is the safety aspect. Any serious damage to a hydride tank (such as one that

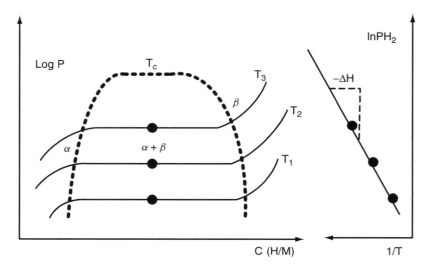

Figure 15 P-C isotherms and Van't Hoff curve for LaNi$_5$ metal hydride.

could be caused by a collision) would not pose a fire hazard because hydrogen would remain in the metal structure.

An overview of hydrogen storage alloys has been discussed by Sandrock[61] from the solid-gas reaction point of view. A number of important properties must be considered in metal hydride storage, including (1) ease of activation; (2) heat transfer rate; (3) kinetics of hydriding and dehydriding; (4) resistance to gaseous impurities; (5) cyclic stability; (6) safety; and, (7) weight and cost. Although metal hydrides can theoretically store large amounts of hydrogen in a safe and compact way, the practical gravimetric hydrogen density is limited to <3 mass%. It is still a challenge to explore the properties of lightweight metals and complex hydrides.

Complex hydrides—MAlH$_4$, MBH$_4$, and N(AlH$_4$)$_2$ (M = Na, Li, K; N = Mg)—are emerging as promising hydrogen storage materials because of their high-potential storage capacity.[30] However, they are generally characterized by irreversible dehydriding or extremely slow hydrogen cycling kinetics. The breakthrough discovery of doping with a few mole percent of Ti-catalyst has enhanced the dehydrogenation kinetics of NaAlH$_4$ at low operating temperatures (<150°C) and was the starting point in reinvestigating these complex hydride systems for hydrogen storage.[62] However, reduced availability of reversible hydrogen (~4−5 wt.%), poor cyclic stability, loss of the catalytic function of Ti-species, necessitates the search for new and efficient complex hydride systems.[63] There are about 234 complex chemical hydrides that have been reported with theoretical hydrogen storage capacity.[64] Table 5 lists complex chemical hydrides under investigation, with their available capacities and operating temperatures.

Table 5   Theoretical Hydrogen Storage Capacities of Complex Hydrides

No.	Complex Chemical Hydride	Theoretical Capacity wt.%	Reversible Capacity wt.%	Operating Temperature °C	Remarks
1.	Ti-doped $NaAlH_4$	7.5	5.5	100–150	High Rehydrogenation pressure, poor cycle life, loss of catalytic activity, less available capacity
2.	Undoped and Ti-doped $LiAlH_4$	10.5	6.3	120–170	Problems with reversibility, and reduced thermodynamic stability
3.	Undoped and doped $LiBH_4$	18.2	9.0	200–400	High operating temperature, rehydrogenation problem, possible borane gas evolution
4.	$Mg(AlH_4)_2$	9.3	6.6	200–250	High operating temperature, thermodynamic stability
5.	$NaBH_4/H_2O$	10.5	9.2	Ambient	Hydrolysis reaction, irreversibility, one-time use
6.	Li_3N ($LiNH_2/LiH$)	11.3	6.5–7.0	255–285	High operating temperature, Possible ammonia evolution
7.	B-H-Li-N	10.0		80–150	Rehydrogenation problem
8.	AlH_3	10.5		150	Ball milling induced decomposition, irreversible
9.	H_3BNH_3	18.3	12.6		Ammonia evolution possibility, irreversible

The hydride complexes such as $NaAlH_4$ and $NaBH_4$ are known to be stable and decompose only at elevated temperatures, often above the melting point of the complex. However, the addition of a few mole concentrations of titanium species to $NaAlH_4$ eases the release of hydrogen at moderate temperatures and ambient pressure.[36] The decomposition of Ti-doped $NaAlH_4$ proceeds in two steps with the total released hydrogen of ~5.5 wt.% at 100° to 150°C, as given in equations (32) and (33).

$$3NaAlH_4 \rightleftarrows Na_3AlH_6 + 2Al + 3H_2 \ (3.72 \ \text{wt.% } H_2) \qquad (32)$$

$$Na_3AlH_6 \rightleftarrows 3NaH + Al + 3/2H_2 \ (1.8 \ \text{wt.% } H_2) \qquad (33)$$

Table 6 New Complex Hydrides and Their Hydrogen Storage Capacity

Serial Number	Complex Hydride	Theoretical Capacity, wt.%	Decomposition Temperature, T_{dec} °C
1.	$LiAlH_2(BH_4)_2$	15.2	
2.	$Mg(BH_4)_2$	14.8	260–280
3.	$NH_4Cl + LiBH_4$	13.6	>ambient
4.	$Ti(BH_4)_3$	12.9	ca. 25
5.	$Fe(BH_4)_3$	11.9	−30 to −10
6.	$Ti(AlH_4)_4$	9.3	−85
7.	$Zr(BH_4)_3$	8.8	<250
8.	$Zn(BH_4)_2$	8.4	85

Following this breakthrough discovery, an effort was initiated in the U.S. DOE Hydrogen program to develop $NaAlH_4$ and related alanates as hydrogen storage materials.[65,66] Another complex hydride, $Mg(AlH_4)_2$ contains 9.6 wt.% of hydrogen that decomposes below 200°C.[67] Some of the new complex hydrides and their theoretical capacities are listed in Table 6.

Borohydride complexes with suitable alkali or alkaline earth metals are a promising class of compounds for hydrogen storage. The hydrogen content can reach values of up to 18 wt.% for $LiBH_4$.[68] The total amount of hydrogen desorbed up to 600°C is 9 wt.%. Mixing $LiBH_4$ with SiO_2 powder lowers the desorption temperature, so that 9 wt.% of hydrogen is liberated below 400°C.[58] Recently, Chen et al. reported a new hydrogen storage system, lithium nitride (Li_3N), which absorbs 11.5 wt.% of hydrogen reversibly.[69,70] The hydrogenation of lithium nitride is a two-step reaction, as shown in equations (34) and (35):

$$Li_3N + 2H_2 \rightleftharpoons Li_2NH + LiH \qquad (34)$$

$$Li_2NH + H_2 \rightleftharpoons LiNH_2 + LiH \qquad (35)$$

Li_3N absorbs 5.74 wt.% of hydrogen for the first step and 11.5 wt.% for second step. Since the hydrogen pressure for the reaction corresponding to the first step is very low (about 0.01 bar at 255°C), only the second step reaction of Li_2NH (lithium imides) with H_2 leads to the reversible storage capacity. According to Chen et al., the plateau pressure for imides hydrogenation is 1 bar at a relatively high temperature of 285°C.[69] However, the temperature of this reaction can be lowered to 220°C with magnesium substitution, although at higher pressures.[71] Further research on this system may lead to additional improvements in operating conditions with improved capacity.

Chemical Hydrogen Storage

The term *chemical hydrogen storage* is used to describe storage technologies in which hydrogen is generated through a chemical reaction. Common reactions involve chemical hydrides with water or alcohols. Typically, these reactions are not easily reversible on board a vehicle. Hence, the *spent fuel,* or byproducts, must be removed from the vehicle and regenerated off board.

Hydrolysis Reactions. Hydrolysis reactions involve the oxidation reaction of chemical hydrides with water to produce hydrogen. The reaction of sodium borohydride has been the most studied to date:

$$NaBH_4 + 2H_2O \rightarrow NaBO_2 + 4H_2 \tag{36}$$

In the first embodiment, slurry of an inert stabilizing liquid protects the hydride from contact with moisture and makes the hydride pumpable. At the point of use, the slurry is mixed with water, and the consequent reaction produces high-purity hydrogen. The reaction can be controlled in an aqueous medium via pH and the use of a catalyst. Although the material hydrogen capacity can be high and the hydrogen release kinetics fast, the borohydride regeneration reaction must take place off board. Regeneration energy requirements cost and life-cycle impacts are key issues currently being investigated.

Millennium Cell has reported that its $NaBH_4$-based Hydrogen on Demand™ system possesses a system gravimetric capacity of about 4 wt.%.[72] Similar to other material approaches, issues include system volume, weight and complexity and water availability.

Another hydrolysis reaction currently being investigated by Safe Hydrogen is the reaction of MgH_2 with water to form $Mg(OH)_2$ and H_2.[73] In this case, particles of MgH_2 are contained in nonaqueous slurry to inhibit premature water reactions when hydrogen generation is not required. Material-based capacities for the MgH_2 slurry reaction with water can be as high as 11 wt.%. However, as with the sodium borohydride approach, water must be carried on board the vehicle in addition to the slurry, and the $Mg(OH)_2$ must be regenerated off board.

New Chemical Approach. A new chemical approach may provide hydrogen generation from ammonia-borane materials by the following reactions:

$$NH_3BH_3 \rightleftarrows NH_2BH_2 + H_2 \rightleftarrows NHBH + H_2 \tag{37}$$

The first reaction, which occurs at less than $120°C$ releases 6.1 wt.% hydrogen, while the second reaction, which occurs at approximately $160°C$, releases 6.5 wt.% hydrogen.[74] Recent studies indicate that hydrogen-release kinetics and selectivity are improved by incorporating ammonia-borane nanosized particles in a mesoporous scaffold.

Carbonaceous Materials for Hydrogen Storage

Carbonaceous materials are attractive candidates for hydrogen storage because of a combination of adsorption ability, high specific surface, pore microstructure, and low mass density. In spite of extensive results available on hydrogen uptake by carbonaceous materials, the actual mechanism of storage still remains a mystery. The interaction may either be based on van der Walls attractive forces (physisorption) or on the overlap of the highest occupied molecular orbital of carbon with occupied electronic wave function of the hydrogen electron, overcoming the activation energy barrier for hydrogen dissociation (chemisorption). The physisorption of hydrogen limits the hydrogen-to-carbon ratio to less than one hydrogen atom per two carbon atoms (i.e., 4.2 mass %). While in chemisorption, the ratio of two hydrogen atoms per one carbon atom is realized, as in the case of polyethylene.[35,75,76] Physisorbed hydrogen has a binding energy normally of the order of 0.1 eV, while chemisorbed hydrogen has C–H covalent bonding, with a binding energy of more than 2 to 3 eV.

Dillon et al. presented the first report on hydrogen storage in carbon nanotubes and triggered a worldwide tide of research on carbonaceous materials.[77] Hydrogen can be physically adsorbed on activated carbon and be "packed" on the surface and inside the carbon structure more densely than if it has just been compressed. The best results achieved with carbon nanotubes to date confirmed by the National Renewable Energy Laboratory is hydrogen storage density corresponding to about 10 percent of the nanotube weight.[78]

Hydrogen can be stored in glass microspheres of approximately 50 μm diameter. The microspheres can be filled with hydrogen by heating them to increase the glass permeability to hydrogen. At room temperature, a pressure of approximately 25 MPa is achieved, resulting in storage density of 14 percent mass fraction and 10 kg H_2/m^3.[79] At 62 MPa, a bed of glass microspheres can store 20 kg H_2/m^3. The release of hydrogen occurs by reheating the spheres to again increase the permeability.

High Surface Area Sorbents and New Materials Concepts

There is a pressing need for the discovery and development of new reversible materials. One new area that may be promising is that of high surface area hydrogen sorbents based on microporous metal-organic frameworks (MOFs). Such materials are synthetic, crystalline, and microporous, and are composed of metal/oxide groups linked together by organic struts. Hydrogen storage capacity at 78K ($-195°C$) has been reported as high as 4 wt.% via an adsorptive mechanism, with a room temperature capacity of approximately 1 wt.%.[80] However, due to the highly porous nature of these materials volumetric capacity may still be a significant issue.

Another class of materials for hydrogen storage may be clathrates, which are primarily hydrogen-bonded H_2O frameworks.[81] Initial studies have indicated that significant amounts of hydrogen molecules can be incorporated into the sII clathrate. Such materials may be particularly viable for off board storage of hydrogen without the need for high pressure or liquid hydrogen tanks.

Other examples of new materials and concepts are conducting polymers. New processes such as sonochemistry may also be applicable to help create unique nano-structures with enhanced properties for hydrogen storage.

4   HYDROGEN UTILIZATION

Today, hydrogen is used primarily in ammonia production, petroleum refinement and the synthesis of methanol. It is also used in the U.S. National Aeronautics and Space Administration's (NASA) space program as fuel for the space shuttles, and in fuel cells that provide heat, electricity, and drinking water for the astronauts. Current thinking suggests that fuel cells are the way to use hydrogen and that the fuel cell industry is the driving force toward a hydrogen economy. This may be true, but it loses sight of other, less costly opportunities. Hydrogen can also be used in internal combustion engines (ICE), turbines, and gas boilers. In many parts of the world, the gas that is used to fuel lights and furnaces is a hydrogen-rich mixture called *town gas,* mainly consisting of hydrogen and methane. In the very near future, hydrogen will be used to fuel vehicles and aircraft, and provide power for our homes and offices.[82] Hydrogen's potential use as a fuel and an energy carrier includes powering vehicles, running turbines or fuel cells to produce electricity, and co-generating heat and electricity for buildings, among others.

4.1   Fuel Cells

Fuel cells are significantly more energy efficient than combustion-based power-generation technologies. A conventional combustion-based power plant typically generates electricity at efficiencies of 33 to 35 percent, while fuel-cell plants can generate electricity at efficiencies of up to 60 percent. When fuel cells are used to generate electricity and heat (co-generation), they can reach efficiencies of up to 85 percent. Internal combustion engines in today's automobiles convert less than 30 percent of the energy in gasoline into power that moves the vehicle. Vehicles using electric motors powered by hydrogen fuel cells are much more energy efficient, utilizing 40 to 60 percent of the fuel's energy. Even fuel-cell vehicles that reform hydrogen from gasoline can use about 40 percent of the energy in the fuel.[83]

A fuel cell can be thought of as an *electrochemical combustor.* Hydrogen is oxidized, some heat is released, and, as in any chemical reaction, electrons change hands (i.e., chemical bonds are broken). However, in a fuel cell, the fuel and oxidant react separately in different regions that are connected to each other by two different conduits for charged particles. This consists of a catalytically activated electrode for the fuel (anode) and the oxidant (cathode), and an electrolyte to conduct ions between the two electrodes. The exchange of electrons among the reagents occurs through an electrical circuit outside the cell. The fuel cell converts chemical potential energy to usable electrical energy in the form of moving electrons. For electrons to journey through the external circuit, they must overcome any electrical barriers, such as impedance, to their transmission in order to do electrical work. Electrochemical reactions for use in fuel cells are purposely chosen so that the amount of electrical work attainable is sufficient to overcome the resistance of electron flow inherent in any circuit, but also to allow the electron flow to carry out useful electrical tasks. The second conduit for charged particles is inside the cell and is called an *electrolyte*. This can be an aqueous or other solution, a solid polymer, or an ion-conducting ceramic. The electrolyte allows particles much more massive than electrons, such as H^+ or OH^-, respectively, for acidic and basic electrolytes or O^{2-} in the case of solid oxide ceramics to pass between the two electrodes. Figure 16 shows an illustration of an acidic electrolyte fuel cell in which H_2 is converted to electricity using O_2 as the oxidant.

This cell requires four chemical and physical processes for this cell to operate:

1. Oxidation of the fuel, gaseous $H_{2(g)}$, at a region of the anode in interfacial contact with the electrolyte:

$$H_{2(g)} \rightarrow 2H^+ + 2e^- \tag{38}$$

2. Physical transport of H^+ from the anode through the electrolyte to the cathode:

$$2H^+ (\text{anode-electrolyte interface}) \rightarrow 2H^+ (\text{electrolyte-cathode interface}) \tag{39}$$

3. Reduction of gaseous O_2, the oxidant, at a region of the cathode in interfacial contact with the electrolyte:

$$\tfrac{1}{2}O_{2(g)} + 2e^- + 2H^+ \rightarrow H_2O \tag{40}$$

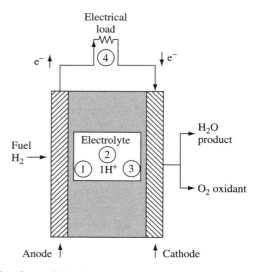

Figure 16   Schematic of an acidic electrolyte hydrogen-oxygen fuel cell illustrating the four essential processes described in the text.

4. Physical transport of electrons from the anode to the cathode through the external circuit:

$$2e^-\text{(anode-electrolyte interface)} \rightarrow 2e^-\text{(cathode-electrolyte interface)} \quad (41)$$

Summarizing equations (38) to (41) one obtains the same overall reaction as direct combustion of hydrogen:

$$H_2 + \tfrac{1}{2}O_2 \rightarrow H_2O \quad (42)$$

However, through the use of the fuel cell, electricity is generated directly. Fuel cells now command great interest as clean energy converters for use in producing electricity for consumers and as the energy source for electric vehicles. This interest is motivated by the potential for high fuel-to-electricity conversion efficiencies and fuel cells run with hydrogen emit only water and waste heat. Since fuel cells operate at lower that typical combustion processes, NO_x emissions are eliminated. Fuel cells are also attractive because of their potential for low maintenance, high reliability, and low noise levels.

Table 7 summarizes several common types of fuel cells under development for stationary electric power systems and vehicle propulsion applications, together with estimates of their fuel-to-electricity conversion efficiencies, assuming that H_2 is the fuel.[84] The ion flowing through the electrolyte may be H^+ of any of several negatively charged species (anions), such as OH^-, CO_3^{2-}, or O^{2-} (the anions flow from the cathode to the anode). Advanced electrolyte systems such as proton-conducting inorganic oxides (i.e. ceramics), may enable fuel-to-electricity efficiencies as high as 70 percent based on H_2 and its low heat value. Because each fuel-cell unit generates approximately 1 volt, fuel-cell systems are composed of stacks of individual fuel cells that are interconnected to produce the desired voltage and power densities for specific application.

Although larger fuel cells (greater than 200 kilowatts) are being commercialized for on-site cogeneration of electricity and steam heat, fuel cells for transportation are in much earlier stages of development. Fuel cells are currently too large, too heavy, and too expensive to produce for a commercial application in powering vehicles. With the resolution of these problems, however, hydrogen-powered fuel cell vehicles will be pollution free and about three times as energy efficient as comparable gasoline-fueled vehicles. There are several configurations of fuel cells, classified by the type of electrolyte used. The most mature technology for near-term use in large vehicles is the phosphoric acid fuel cell. The proton-exchange membrane fuel cell is a prime candidate for mid-term use in several areas, including automobiles. Solid oxide fuel cells are being developed for longer-term utility applications.[85] Descriptions of the common fuel cells from Table 7 are presented as follows.

Table 7 Summary of Common Fuel-cell Technologies (Adapted from Ref. 84)

Fuel Cell	Anode Reaction	Electrolyte	Transfer Ion[a]	Cathode Reaction	Operating Temp. (°C)	Conditions Pressure (atm)	H_2–to-Electrical Efficiency, %
Proton exchange membrane (PEM)[b]	$H_2 \rightarrow 2H^+ + 2e^-$	Solid polymer	H^+	$2H^+ + \frac{1}{2}O_2 + 2e^- \rightarrow H_2O$	80–100	1–8[c]	36–38
Phosphoric acid (PAFC)	$H_2 \rightarrow 2H^+ + 2e^-$	Phosphoric acid	H^+	$2H^+ + \frac{1}{2}O_2 + 2e^- \rightarrow H_2O$	150–250	1–8[c]	40
Alkaline (AFC)	$H_2 \rightarrow 2H^+ + 2e^-$	Aqueous base	OH^-	$H_2O + \frac{1}{2}O_2 + 2e^- \rightarrow 2OH^-$	80–250	1–10[c]	50 to >60
Molten carbonate (MCFC)	$H_2 + CO_3^{2-} \rightarrow H_2O + CO_2 + 2e^-$	Molten salt (metal carbonate)	CO_3^{2-}	$CO_2 + \frac{1}{2}O_2 + 2e^- \rightarrow CO_3^{2-}$	600–700	1–10	50–55
Solid oxide (SOFC)	$H_2 + O^{2-} \rightarrow H_2O + 2e^-$	Solid ceramic oxide	O^{2-}	$2H^+ + \frac{1}{2}O_2 + 2e^- \rightarrow H_2O$	800–1,000	1	50–55

[a]Positively charged ions (cations) travel thought the electrolyte from the anode to the cathode; negatively charged ions (anions) transverse the electrolyte in the opposite direction.
[b]Also called Polymer Electrolyte Fuel Cell.
[c]Pressure must be sufficiently high to prevent boiling of water in PEM or aqueous electrolytes in PAFCs and AFCs.

Phosphoric Acid Fuel Cells

A phosphoric acid fuel cell (PAFC) consists of an anode and a cathode made of finely dispersed platinum catalyst on carbon paper, and a silicon carbide matrix that holds the phosphoric acid electrolyte. PAFCs produce a cell voltage of 0.66 volts at atmospheric pressure and 200°C, and a current density of 240 milliampere per square centimeter (mA/cm^2). Overall fuel-to-electricity energy-conversion efficiency is about 40 percent. PAFCs are the most advanced of the fuel-cell designs, and are being commercialized for stationary power applications and for demonstrations in larger fleet vehicles, such as buses. The power density of a PAFC is too low for use in an automobile, however, and it cannot generate power at room temperature. Because of these limitations, the optimum use of PAFCs is in steady-state operating modes. Researchers are studying other fuel-cell alternatives for vehicle applications.

Proton Exchange Membrane Fuel Cells

The proton exchange membrane (PEM) fuel cell uses a fluorocarbon ion exchange with a polymeric membrane as the electrolyte. The hydrogen proton migrates across the membrane, and water is evolved at the cathode. The PEM operates at a relatively low temperature of about 80°C and can start up from ambient temperature at partial load. These characteristics, plus its high-power density, make the PEM cell more adaptable to automobile use than the PAFC. Current densities of up to 4 A/cm^2 have been reported for single PEM cells. An assembly of PEM cells has not been able to achieve this level because at high current densities localized overheating limits the attainable density to about 1 A/cm^2. As research overcomes this problem, higher current densities will allow the weight and volume of a PEM fuel cell to be more practical for vehicle use.

Solid Electrolyte Fuel Cells

Solid oxide fuel cells (SOFC) currently under development use a thin layer of zirconium oxide as a solid electrolyte, a lanthanum manganite cathode, and a nickel-zirconia anode. When heated to about 1,000°C, the oxide becomes a suitable conductor of oxygen ions but not electrons. A tubular arrangement of the cathode, anode, and electrolyte is the most advanced of the SOFC designs; 20-kilowatt demonstration units have been installed in Japanese utilities. A planar configuration consists of alternating flat plates of a trilayer containing an anode, an electrolyte, and a cathode. A monolithic configuration adds a layer of anode and cathode material corrugated on either side of the trilayers to form flow channels for the fuel and air streams. Planar SOFCs are easier to fabricate than the monolithic configuration, which is co-sintered into a solid, ceramic structure, but monolithic configurations have the highest-power density of the designs. All SOFC designs have fewer components and ultimately may need less maintenance and be less expensive than other fuel-cell types.

Fuel cells are but one of many opportunities for hydrogen energy utilization. As described next, internal combustion engines and hydrogen burners are relatively low-cost hydrogen utilization technologies.

4.2 Internal Combustion Engines

Hydrogen use in an internal combustion engine (ICE) was demonstrated more than 100 years ago. Hydrogen-fueled ICE offer the potential of no carbon and very low nitrogen oxide emissions, combined with high thermal efficiency. To be competitive and cost-effective, however, key problems must be solved in engine combustion, fuel delivery, and practical storage. The primary research goal is to develop an optimized hydrogen IC engine with about 80 miles per gallon equivalent performance in an ultra-low-emission vehicle. Fuel efficiencies of 80 to 90 miles per gallon energy equivalent have been realized in simulations using a hybrid hydrogenelectric vehicle. The research challenge is to achieve high efficiency and low emissions while overcoming the problems of preignition and flashback that have been common with hydrogen fuel in the past. *Flashback* is the improperly timed explosion of the fuel and air mixture that occurs when the exhaust valve of the ICE is open. This risk is more significant with hydrogen fuel than hydrocarbon fuels because hydrogen's flame speed is 2 to 10 times greater than that of hydrocarbons. Two key areas of investigation are the fuel delivery system and the ignition system. In a carburetion system, premixing creates a lean, homogeneous charge that keeps nitrogen oxide emissions low. A fuel-injected system better prevents preignition of the fuel-air mixture and flashback. These engines can be used for both transportation and stationary power applications. Researchers are studying direct power from an ICE fueled by hydrogen or mixed fuels (such as hydrogen-methane) and hybrid power systems, where an ICE operating at a single speed and load runs an electric motor.

Research is also focused on reducing nitrogen oxide emissions in fuel-injection systems by diluting the intake air charge in a direct-injection ICE. Dilution can be accomplished by recirculating exhaust gases or by scavenging. These techniques work by reducing the flame temperature and oxygen availability of the hydrogen/oxidizer mixture. This subsequently reduces the formation of nitrogen oxides, which is highly sensitive to temperature. It is anticipated that hydrogen ICEs will provide a high-volume usage of hydrogen prior to the low-cost mass production of fuel cells for transportation. In this light, the use of hydrogen or hydrogen-methane mixtures in ICEs could result in an acceleration of hydrogen demand into the nearer-term economy, particularly in terms of hydrogen production and delivery. Several programs are currently underway to define baseline emissions and operational ranges for hydrogen and hydrogen-methane mixtures. Examples of these types of programs are studies being conducted by the U.S. National Energy Technology Laboratory's (NETL) Office of Science and Engineering Research (OSER) on lubricant life in hydrogen ICEs, injector performance, and ignition systems.[86]

4.3   Hydrogen Burner Turbines

Research is focusing on the development of a safe and environmentally benign hydrogen burner that can generate electricity for utilities and provide heat to industry and homes. Burning hydrogen eliminates most emissions that come from carbon-based fuels, including carbon dioxide and carbon monoxide. The burning of any fuel in air, however, produces some amount of nitrogen oxides, and burner research is focusing on eliminating these emissions from hydrogen combustion. One way to do this is to remove nitrogen from the fuel mix completely, by burning pure hydrogen and pure oxygen derived directly from the electrolysis process. This is an expensive alternative, however, and researchers are looking at more cost-effective methods. Nitrogen oxide emissions can be minimized by reducing the peak combustion temperature and the time spent at the peak temperature. Typical thermal efforts reduce the peak temperature by recirculating cooler inert gases through the combustion process or injecting steam. Nitrogen oxides can also be reduced to essentially zero by premixing the fuel and oxidizer to reduce the amount of fuel in proportion to the oxidizer—a *lean* mixture. A sufficiently lean mixture can reduce the combustion temperatures to 1,400°C to 1,500°C, although it can also increase the occurrence of flashback. Researchers are investigating the combustion fluid dynamics required to completely oxidize hydrocarbon and hydrogen fuels. Because the momentum flux of the oxidizer (air) is the primary variable in resolving these problems, an improved hydrogen burner will also work efficiently with natural gas and liquid petroleum gas. This flexibility should accelerate the utilization of hydrogen by facilitating the use of hybrid fuels.[87]

5   HYDROGEN SAFETY

Hydrogen has been safely used for a long time in industrial and aerospace applications. Through this experience, a great deal of relevant knowledge exists. However, in the preliminary stages of a hydrogen economy, great care must be taken to assure a high degree of safety in all hydrogen applications, because a loss in public confidence could have a significant impact on future developments.

5.1   The Nature of Hydrogen

Hydrogen is less flammable than gasoline. The self-ignition temperature of hydrogen is 585°C. The self-ignition temperature of gasoline varies from 228° to 501°C, depending on the grade (Table 8). Hydrogen disperses quickly and, being the lightest element (15 times lighter than air), it rises and spreads out quickly in the atmosphere. So when a leak occurs, the hydrogen gas quickly becomes so sparse that it cannot burn. Even when ignited, hydrogen burns upward, and is quickly consumed. By contrast, materials such as gasoline or natural gas are heavier than air, and will not disperse, remaining a flammable threat for

Table 8   Summary of Safety Statistics for Hydrogen and Other Fuels (Ref. 89.)

Characteristic	Hydrogen	Natural Gas	Gasoline
Lower heating value kJ/g	120	50	44.5
Self-ignition temperature ($^{\circ}$C)	585	540	228–501
Flame temperature ($^{\circ}$C)	2,045	1,875	2,200
Flammability limits in air (vol%)	4–75	5.3–15	1.0–7.6
Minimum ignition energy in air (μJ)	20	290	240
Detonability limits in air (vol%)	18–59	6.3–13.5	1.1–3.3
Theoretical explosive energy (kg TNT/m3 gas)	2.02	7.03	44.22
Diffusion coefficient in air (cm^2/s)	.61	.16	.05

a longer period of time. Hydrogen is a nontoxic, naturally occurring element in the atmosphere. By contrast, all fossil fuels are poisonous to humans. Hydrogen combustion produces only water. Compared with the toxic compounds (carbon monoxide, nitrogen oxides, and hydrogen sulfide) produced by petroleum fuels, the products of hydrogen burning are much safer. Hydrogen can be stored safely in gaseous, liquid or solid state form (see hydrogen storage section). Tanks currently in use for storage of compressed hydrogen (similar to compressed natural gas tanks) have survived intact through testing by various means, such as bullets, fires, and shocks.[88]

Other properties of hydrogen necessitate special considerations when handling. Hydrogen consists of small molecules, which require special qualities in materials used in storage and transportation. Hydrogen creates flammable and explosive mixtures of air over a broad spectrum (Table 8).[89] These mixtures need very little energy to ignite. Ventilation is therefore an important factor in areas where hydrogen is used.

5.2   How to Handle Hydrogen

The wide flammability ranges of hydrogen imply that a mixture of hydrogen and air might ignite more easily than other fuels. Consequently, the following precautions must be adhered to:

- Hydrogen should not be mixed with air.
- Contact of hydrogen with potential ignition sources should be prevented.
- Purging hydrogen systems should be performed with an inert gas such as nitrogen.
- Venting hydrogen should be done according to standards and regulation.
- Because the hydrogen flame is invisible, special flame detectors are required.

Hydrogen should be handled with special care in confined, unvented areas. Various safety assessments (safety codes and standards) are available or under development to serve as a guide in setting up and designing hydrogen systems.[90]

Codes and Standards

The subject of codes and standards is covered in a different section of this handbook. However, it is important to note that the Hydrogen, Fuel Cells, and Infrastructure Technologies (HFCIT) Program of the U.S. Department of Energy (DOE) and the National Renewable Energy Laboratory (NREL) are developing hydrogen codes and standards to expedite the future construction of a hydrogen infrastructure. HFCIT has developed a Web-based bibliographic database that is intended to provide easy public access to a wide range of hydrogen safety aspects. The database includes references related to the following topics:[91]

- Hydrogen properties and behavior
- Safe operating and handling procedures
- Leaks, dispersion, and flammable vapor cloud formation
- Embrittlement and other effects on material properties
- Fuel cells and other energy conversion technologies
- Sensors, tracers, and leak detection technologies
- Accidents and incidents involving hydrogen

Also, the National Fire Protection Association (NFPA) has incorporated hydrogen safety requirements in its family of codes and standards.[92] In Europe, a HySafe Network of Excellence for Hydrogen Safety has been formed. The network is composed of 24 partners from 12 European countries and Canada, representing private industries, universities, and research institutions.[93] The Web site www.hysafe.net offers a wealth of information on hydrogen safety. The New Energy and Industrial Technology Development Organization (NEDO) in Japan is pursuing a large number of projects on the safety of hydrogen infrastructure and building frames in case of hydrogen explosion and earthquakes.[94–96]

Other organizations are also involved in new standards activities. The National Hydrogen Association (NHA) has created Codes and Standards Working Groups on topics such as hydride storage, electolyzers for home use, transportation infrastructure issues, and maritime applications. The Society of Automotive Engineers, through a Fuel Cell Standards Forum Safety Task Force, is collaborating with NHA on the transportation issues. The International Organization for Standardization (ISO) level in ISO Technical Committee 197 (Hydrogen Technologies) is actively pursuing the development of codes and standards with input from national organizations.[97]

6 CONCLUSIONS

Fossil fuels, electricity, biomass, and sunlight are four potential resources to use in H_2 production. So far, hydrogen has been produced principally from methane (a depleting energy resource) using steam reforming. Although several possibilities exist for hydrogen production, solar-based hydrogen would be desirable.

Further, hydrogen represents a good storage medium of solar energy. Producing hydrogen from water using solar energy appears to be an attractive step toward this approach. However, relatively few water-based solar hydrogen-producing systems are currently available: thermochemical cycle, photoelectrochemical system, photochemical process, and solar assisted electrolysis.

Even though thermochemical systems have high theoretical limits, they exhibit problems with materials and separation at high temperatures. Photochemical and photoelectrochemical systems are currently at a very early stage of development. The difficulties that still need to be addressed with photoelectrochemical systems are semiconductor stability, efficient light absorption and interfacial kinetics. Currently, biophotolysis processes demonstrate very low solar conversion efficiencies, and they can be sustained only for short periods of time. Among the various hydrogen production methods, water electrolysis is the only developed nonpolluting technology. Electrolysis efficiency of 85 to 95 percent is currently possible.

Hydrogen storage is essential, especially for the on-board vehicular applications that lead to a hydrogen based economy. Various hydrogen storage methods have been presented in this chapter with respect to their physical and chemical phenomena. Currently, none of the storage methods are mature enough to address all the technological barriers and targets of the U.S. DOE's FreedomCAR goals and require additional basic and applied research.

Some technologies are currently available for the practical and cost-effective utilization of hydrogen as an energy carrier. Additional technologies need to be developed as hydrogen production, transport, and storage capabilities become integrated into the energy economy.

Clean energy and a healthy environment are the concerns of everyone. Aggressive improvement in energy efficiency, along with well-thought-out and executed transitional strategies, are essential to enable the growth of hydrogen utilization and the development of technologies, markets, and infrastructure to support a green hydrogen economy.

REFERENCES

1. A. Raissi, "Technoeconomic Analysis of Area II Hydrogen Production—Part 1," Proceedings of the 2001 U.S. DOE Hydrogen Program Review, NREL/CP/570-30535, 2006.
2. R. Minet and K. Desai, "Cost-Effective Methods for Hydrogen Production," *International Journal for Hydrogen Energy*, **8**, p. 285–290 (1983).
3. S. Sherif, T.N. Veziroglu, and F. Barbir, "Hydrogen Energy Systems," in J. G. Webster, (ed.), *Wiley Encyclopedia of Electrical and Electronics Engineering*, vol. 9, John Wiley, New York, 1999, pp. 370–402.
4. B. Gaudernack and Lynum, S. "Hydrogen from Natural Gas without Release of CO_2 to the Atmosphere," Proceedings of the 11th World Hydrogen Energy Conference: Hydrogen Energy Progress, 511–523, Germany, 1996.

5. D. W. Kirk and A. E. Ledas, "Precipitate Formation during Sea-Water Electrolysis," *International Journal of Hydrogen Energy*, **7**, 925–932 (1982).

6. L. Basye and S. Swaminathan. "Hydrogen Production Costs—A Survey," U.S. Department of Energy, DOE/GO/10170-T18, 1997.

7. H. Audus, O. Kaarstad and M. Kowal. "Decarbonization of Fossil Fuels: Hydrogen as an Energy Carrier," Proceedings of the 11th World Hydrogen Energy Conference: Hydrogen Energy Progress, 1, 525–534 (1996).

8. C. E. G. Padro and V. Putsche, "Survey of the Economics of Hydrogen Technologies," National Renewable Energy Laboratory, NREL/TP-570-27079, 1999.

9. S. Leiby, *"Options for Refinery Hydrogen,"* Private report by The Process Economics Program, Report No. 212, SRI International, Menlo Park, CA, 1994.

10. T. N. Veziroglu, "Hydrogen Technology for Energy Needs of Human Settlements," *International Journal of Hydrogen Energy*, **12**(2), 99–129 (1987).

11. P.L. Spath, M.K. Mann, and W.A. Amos, "Update of Hydrogen from Biomass: Determination of the Delivered Cost of Hydrogen," NREL/MP-510-33112, 2000.

12. D. Dayton, "A Review of the Literature on Catalytic Biomass Tar Destruction: Milestone Completion Report," NREL/TP-510-32815, December 2002.

13. L. Devi, K. J. Ptasinski, and F. J. J. G. Janssen, "A Review of the Primary Measures for Tar Elimination in Biomass Gasification Process," *Biomass and Bioenergy*, **24**, 125–140, (2003).

14. M. K. Mann, "Technical and Economic Analysis of Hydrogen Production via Indirectly Heated Gasification and Pyrolysis," Proceedings of the 1995 USDOE Hydrogen Program Review, NREL/CP-430-20036, 1, 205–236 (1995).

15. T. A. Milne, C. C. Elam, and R. J. Evans, "Hydrogen from Biomass: State of the Art and Research Challenges," NREL IEA/H$_2$/TR-02/001 (2001).

16. H. Wendt, "Electrochemical Hydrogen Technologies: Electrochemical Production and Combustion of Hydrogen," Elsevier, New York, 1990.

17. S. M. El-Haggar and M. Khalil, "Parametric Study of Solar Hydrogen Production from Saline Water Electrolysis," *International Journal of Environment and Pollution*, **8**, 164–173 (1997).

18. S. Zhuiykov, "Research on Proton-Conducting Materials in the Hydrogen Partial Pressure Sensors," *Herald Kiev Polytechnic Institute Instrumentation*, **19**, 7–9 (1989).

19. M. S. Casper. *"Hydrogen Manufacture by Electrolysis, Thermal Decomposition and Unusual Techniques,"* Noyes Data Corporation, New Jersey, 1978.

20. A. Weidenkaff, A. Reller, A. Wokaun, and A. Steinfeld, "Thermogravimetric Analysis of the Zno/Zn Water Splitting Cycle," *Thermochimica Acta*, **359**, 69–75 (2000).

21. M. Sakurai, E. Bilgen, A. Tsustsumi, and Yoshida, K. "Adiabatic UT-3 Thermochemical Process for Hydrogen Production," *International Journal of Hydrogen Energy*, **21**, 865–870 (1996a).

22. T. Nakayama, H. Yoshioka, H. Furutani, H. Kameyama, and K. Yoshida, "Mascot—A Bench Scale Plant for Producing Hydrogen by the UT-3 Thermochemical Decomposition Cycle," *International Journal of Hydrogen Energy*, **9**, 187–190 (1984).

23. M. Sakurai, N. Miyake, A. Tsutsumi, and Yoshida, K. "Analysis of a Reaction Mechanism in the UT-3 Thermochemical Hydrogen Production Cycle," *International Journal of Hydrogen Energy*, **21**, 871–875 (1996b).

24. R. Amir, S. Shizaki, K. Tamamoto, T. Kabe, and H. Kameyama. "Design Development of Iron Solid Reactants in the UT-3 Water Decomposition Cycle Based on Ceramic Support Materials," *International Journal of Hydrogen Energy*, **18** 283–286 (1993)

25. M. Sakurai, M. Aihara, N. Miyake, A. Tsutsumi, and K. Yoshida, "Test of One-Loop Flow Scheme for the UT-3 Thermochemical Hydrogen Production Process," *International Journal of Hydrogen Energy*, **17**, 587–592 (1992).

26. M. Sakurai, H. Nakajima, K. Onuki, K. Ikenoya, and S. Shimizu, "Preliminary Process Analysis for the Closed Cycle Operation of the Iodine-Sulfur Thermochemical Hydrogen Production Process," *International Journal of Hydrogen Energy*, **24**, 603–612 (1999).

27. K. R. Schultz, L. C. Brown, and C. J. Hamilton, Production of Hydrogen by Nuclear Energy: The Enabling Technology for the Hydrogen Economy, American Nuclear Energy Symposium, Miami, Florida, October 16–18, 2002.

28. J. R. Bolton, S. J. Strickler, and J. S. Connolly. "Limiting and Realizable Efficiencies of Solar Photolysis of Water," *Nature*, **316**, 495–500.

29. Argonne National Laboratory, Report of the Basic Energy Science Workshop on Hydrogen Production, Storage and Use, Rockville, Maryland, May 13–15, 2003.

30. W. Grochala and P. P. Edwards, "Thermal Decomposition of the Non-Interstitial Hydrides for the Storage and Production of Hydrogen," *Chemical Reviews*, **104**, 1283–1315 (2004).

31. F. E. Pinkerton and B. G. Wicke, *"Bottling the Hydrogen Genie," "The Industrial Physicist"*, 20–25, (February/March 2004).

32. L. Schlapbach, and A. Zuttel, "Hydrogen-Storage Materials for Mobile Applications," *Nature*, **414**(15), 353 (November 2001).

33. A. K. Shukla, "Fuelling Future Cars," *Journal of the Indian Institute of Science*, **85**, 51–65, (March–April, 2005).

34. A. Zuttel, "Hydrogen Storage Methods," *Naturwissenschaften*, **91**, 157–172 (2004).

35. U.S. DOE Energy Efficiency and Renewable Energy (EERE) (2006a), Hydrogen and Fuel Cells, http://www.eere.energy.gov/hydrogenandfuelcells/storage/doe_rd.html.

36. U.S. DOE Energy Efficiency and Renewable Energy (EERE) (2006b), Hydrogen and Fuel Cells, http://www.eere.energy.gov/hydrogenandfuelcells/pdfs/freedomcar_targets_explanations.pdf.

37. D. Chandra, J. J. Reilly, and R. Chellappa. "Metal Hydrides for Vehicular Applications: The State of the Art," *Journal of Materials* (February 26–32, 2006).

38. M. Fichtner, "Nanotechnological Aspects in Materials for Hydrogen Storage," *Advanced Engineering Materials*, **7**(6), 443–455 (2005).

39. M. V. C. Sastry, B. Viswanathan, and S. Srinivasan Murthy, "Metal Hydrides, Fundamentals and Applications," Springer-Verlag, Narosa Publishing, Berlin, Heidelberg, New York, 1998.

40. L. Zhou, "Progress and Problems in the Hydrogen Storage Methods," *Renewable and Sustainable Energy Reviews*, **9**, 395–408 (2005).

41. U.S. DOE Energy Efficiency and Renewable Energy (EERE) (2006c), Hydrogen and Fuel Cells, http://www.eere.energy.gov/hydrogenandfuelcells/posture_plan04.html.

42. J. Ko, K. Newell, B. Geving, and W. Dubno, U.S. DOE Hydrogen Program, FY 2005 Progress Report, http://www.hydrogen.energy.gov/pdfs/progress05/vi_e_1_ko.pdf, Accessed 2005.

43. Y. Fukai, *The Metal-Hydrogen System, Basic Bulk Properties*, 1st ed., Springer-Verlag Publications, Berlin, Heidelberg, 1993.

44. G. G. Libowitz and A. J. Maeland, "Hydrides of Rare Earths," in K. A. J. Gschneidner and L. Eyring (Eds.) *Handbook on the Physics and Chemistry of Rare Earths*, North Holland Publishing, Amsterdam, 1979.

45. W. M. Muller, J. P. Blackledge, and G. G. Libowitz, *Metal Hydrides*, Academic Press, New York and London, 1968.

46. K. J. Gross, *"Intermetallic Materials for Hydrogen Storage,"* Ph.D. Thesis, University of Fribourg, Fribourg, Switzerland, 1998.

47. S. S. S. Raman, *Synthesis and Characterization of Some Mg Bearing Hydrogen Storage (Hydride) Materials*, Ph.D. Thesis, Banaras Hindu University, Varanasi, India, 2006.

48. J. J. Reilly, *Metal Hydrides as Hydrogen Storage and their Applications*, CRC Press, Cleveland, Ohio, 2, 13 (1977).

49. G. D. Sandrock, *"The Metallurgy and Production of Rechargeable Hydrides,"* in Anderson A. F. and Maeland A. J. (Eds.) "Hydrides for Energy Storage", Pergamon, Oxford, 1978, p. 353

50. L. Schlapbach, A. Seiler, F. Stucki, and H. C. Siegman, "Surface Effects and the Formation of Metal Hydrides," *Journal of the Less Common Metals*, **T3**, 145 (1980).

51. B. L. Shaw, *Inorganic Hydrides*, Pergamon Press, United Kingdom, 1967.

52. J.J. Reilly, and R.H. Wiswall Jr., "Formation and Properties of Iron Titanium Hydride," *Inorganic Chemistry*, **13**, 218 (1974).

53. J. H. N. Van Vucht, F.A. Kuijpers, and H. C. A. M. Burning, *"AB_5 Intermetallic Hydrides,"* Philips Research Report, 1970, pp. 133.

54. J. J. Reilly and R. H. Wiswall Jr., "Reaction of Hydrogen with Alloys of Magnesium and Nickel and the Formation of Mg_2NiH_4," *Inorganic Chemistry*, **7**, 225 (1968).

55. P. Selvam, "Energy and Environment—An All Time Research," *International Journal of Hydrogen Energy*, **16**(1), 35–45 (1991).

56. S. S. S. Raman and O. N. Srivastava, "Hydrogenation Behavior of the New Composite Storage Materials, Mg-Xwt.% Cfmmni$_5$," *Journal of Alloys And Compounds*, **241**, 167–174 (1996).

57. A. Zuttel, P. Wenger, S. Rentsch, P. Sudan, P. Mauron, and C. Emmenegger, "LibH$_4$: A New Hydrogen Storage Material," *Journal of Power Sources*, **118**(1–2) 1–7 (2003).

58. J. J. Didisheim, P. Zolliker, K. Yvon, P. Fischer, J. Schefer, M. Gubelmann, and A. Williams, "Dimagnesium Iron (II) Hydride Mg_2FeH_6 Containing Octohedral $[Feh_6]^{4-}$ Anions," *Inorganic Chemistry*, **23**(13), 1953–1957 (1984).

59. L. Schlapbach, I. Anderson, and J. P. Burger, *"Heats of Formations of Metal Hydrides,"* in K. H. J. Buschow, (ed.), "Material Science and Technology", 63, VCH Mbh Publishing, York, UK, 1988.

60. G. Sandrock, "A Panoramic Overview of Hydrogen Storage Alloys from a Gas Reaction Point of View," *Journal of Alloys Compounds*, 293–295, **877** (1999).

61. B. Bogdanovic and M. Shwickardi. "Ti-Doped Alkali Metal Aluminum Hydrides as Potential Novel Reversible Hydrogen Storage Materials," *Journal of Alloys Compounds*, **1**, 253–254 (1997).

62. F. Schuth, B. Bogdanovic, and M. Felderhoff, "Light Metal Hydrides and Complex Hydrides for Hydrogen Storage," *Chemical Communication*, **20**, 2249–2258 (2004).

63. Sandia National Laboratory Hydpark, http://hydpark.ca.sandia.gov. Accessed 2006.

64. C. M. Jensen and R. A. Zidan, "Sodium Alanates for Reversible Hydrogen Storage," Proceedings of the 1998 U.S. Hydrogen Program Review, Alexandria, VA, 449, 1998.

65. C. M. Jensen and K. J. Gross, "Development of Catalytically Enhanced Sodium Aluminum Hydride as a Hydrogen-Storage Material," *Applied Physics A*, **72**, 213, (2001).

66. M. Fichtner, "Synthesis and Structure of Magnesium Alanates and Two Solven Adducts," *Journal of Alloys and Compounds*, **345**(1–2), 286–296 (2002).

67. E. M. Fedneva, V. L. Alpatova, and V. I. Mikheeva, "$LIBH_4$ Complex Hydride Materials," *Russian Journal of Inorganic Chemistry*, **9**, 826, (1964).

68. P. Chen, Z. Xiong, J. Luo, J. Lin, and K. L. Tan. "Interaction of Hydrogen with Metal Nitrides and Imides," *Nature*, **420** (2), 302 (2002).

69. P. Chen, Z. Xiong, J. Luo, J. Lin, and K.L. Tan. "Interaction between Lithium Amide and Lithium Hydride," *Journal of Physical Chemistry B*, **107** 10967 (2003).

70. W. F. Luo, "$Linh_2$-Mgh_2: A Viable Hydrogen Storage System," *Journal of Alloys Compounds*, **381**, 284–287 (2004).

71. Stanford University Global Climate and Energy Project, http://gcep.stanford.edu/pdfs/hydrogen_workshop/wu.pdf. 2003.

72. U.S. DOE Energy Efficiency and Renewable Energy (EERE) (2006d), Hydrogen and Fuel Cells http://www1.eere.energy.gov/hydrogenandfuelcells/pdfs/review04/st_6_mcclaine.pdf.

73. U.S. DOE Energy Efficiency and Renewable Energy (EERE) (2006e), Hydrogen and Fuel Cells, http://www1.eere.energy.gov/hydrogenandfuelcells/pdfs/review04/st_p2_autrey_04.pdf.

74. B. Viswanathan, M. Sankaran, and M. A. Schibioh, "Carbon Nanomaterials: Are They Appropriate Candidates for Hydrogen Storage?" *Bulletin of the Catalysis Society of India*, **2**, 12 (2003).

75. M. G. Nijkamp, J. E. M. J. Raaymakers, A. J. Van Dillen, and K. P. De Jong, "Hydrogen Storage by using Physisorption Materials Demand," *Applied Physics A* **72**, 619 (2001).

76. A. C. Dillon, K. M. Jones, T. A. Bekkedahl, C. H. Kiang, D. S. Bethune, and M. J. Heben, "Storage of Hydrogen in Single-Walled Carbon Nanotubes," *Nature* (London), **386**(6623), 377–379 (1997).

77. P. M. F. J. Costa, K. S. Coleman, and Green, M. L. J. "Influence of Catalyst Metal Particles on the Hydrogen Sorption of Single-Wall Carbon Nanotube Materials," *Nanotechnology*, **16**, 512 (2005).

78. Y. Kojima, Y. Kawai, A. Koiwai, N. Suzuki, T. Haga, T. Hioki, and K. Tange, "IR Characterizations of Lithium Imide and Amide," *Journal of Alloys Compounds*, **395**(1–2) 236–239 (2005).

79. J. L. C. Rowsell, E. C. Spencer, J. Eckert, J. A. K. Howard, and O. M. Yaghi, "Gas Adsorption Sites in a Large Pore Metal-Organic Framework," *Science* **309**, 1350–1354 (2005).

80. F. Schuth, "Technology: Hydrogen and Hydrates," *Nature*, **434**(7034), 712–713, (April 2005).

81. M. Momirlan and T.N. Veziroglu, "The Properties of Hydrogen as Fuel for Tomorrow as Sustainable Energy System for a Cleaner Planet," *International Journal of Hydrogen Energy*, **30**, 795–802 (2005).

82. J. B. O'Sullivan, "Hydrogen Technical Advisory Panel Report and Fuel Cell Development Status." National Research Council Committee May 11, 1999.

83. National Academies Press, "The Hydrogen Economy: Opportunities, Costs, Barriers, and Needs," http://www.nap.edu, 2004.

84. U.S. DOE (2005), Hydrogen from Coal Program—Research, Development, and Demonstration Plan, 2004–2015, September, http://www.netl.doe.gov/technologies/hydrogen_clean_fuels/refshelf/myrddp.html.

85. C.Y. Termaath, E.G. Skolnik, R.W. Schefer, and J.O. Keller, "Emissions Reduction Benefits from Hydrogen Addition to Midsize Gas Turbine Feedstocks," *International Journal of Hydrogen Energy* **31**, 1147–1158 (2006).

86. A. Bain "The Freedom Element—Living with Hydrogen," Blue Note Publications, Cocoa Beach, Florida, 2001.

87. Bellona, "Hydrogen—Status and Possibilities," *Bellona Report*, **4**, (2004), 193.71.199.52/en/energy/hydrogen/report_6-2002/22966.html. Accessed February 1, 2002.

88. National Aeronautics and Space Administration (NASA), "Safety Standards for Hydrogen Safety and Hydrogen Systems," http://www.nasa.gov Accessed 2006.

89. B. Kinzey, P. Davis, and A. Ruiz, "The Hydrogen Safety Program of the U.S. Department of Energy," Proceedings of the 16th World Hydrogen Energy Conference (WHEC), June 13–16, Lyon, France, 2006.

90. J. Ohi, "Hydrogen Codes and Standards: An Overview of the U.S. DOE Activities," Proceedings of the 16th World Hydrogen Energy Conference (WHEC), June 13–16, 2006, Lyon, France.

91. T. Jordan, "Hysafe—The Network of Excellence for Hydrogen Safety" Proceedings of the 16th World Hydrogen Energy Conference (WHEC), June 13–16, Lyon, France, 2006.

92. NEDO, "Research on the Safety of Hydrogen Infrastructure and Building Frame Structure against Hydrogen Explosion and Earthquake," NEDO-03002978-0 and NEDO-04000481-1, http://www.nedo.go.jp/ Accessed 2005.

93. H. Miyahara, K. Kubo, M. Saito, Y. Suwa, K. Yonezawa, K. Naganuma, and K. Imoto, "Research on the Safety of Hydrogen Infrastructure and Building Frame Structure against Hydrogen Explosion and Earthquake," Proceedings of the 15th World Hydrogen Energy Conference (WHEC), Yokohama, Japan, 2006.

94. Y. Suwa, H. Miyahara, K. Kubo, K. Yonezawa, Y. Ono, and K. Mikoda, "Design of Safe Hydrogen Refueling Stations against Gas-Leakage, Explosion and Accidental Automobile Collision," Proceedings of the 16th World Hydrogen Energy Conference (WHEC) June 13–16, 2006, Lyon, France.

95. The Hydrogen Community http://www.hydrogensociety.net, Accessed February 1, 2007.

96. T. Bak, J. Nowotny, M. Rekas, and C. C. Sorrell, "Photo-Electrochemical Hydrogen Generation from Water Using Solar Energy. Materials-Related Aspects," *International Journal of Hydrogen Energy*, **27**, 991–1022 (2002).

97. J.O'M, Bockis, T. N., Veziroglu, and D. Smith, "Solar Hydrogen Energy: The Power to Save the Earth," Macdonald Optima Publications, London, UK, 1991.

98. S. Dunn, "Hydrogen Futures: Toward a Sustainable Energy System," *International Journal of Hydrogen Energy*, **27**, 235–64 (2002).

99. R. Griessen, and T. Riesterer, *"Hydrogen in Intermetallic Compounds,"* in I. L. Schlapbach (Ed.), *Hydrogen in Intermetallic Compounds*, New York, 1988, p. 219.

100. M. Lordgooei, S. A. Sherif, and T. N. Veziroglu, *"Analysis of Liquid Hydrogen Boil-Off Losses,"* "Energy Environment Progress" I, D 241–266 (1991).

101. J. M. Ogden, "Hydrogen: The Fuel of the Future?" *Physics Today*, **55**(4):69–75 (2002).

102. C. Read, J. Petrovic, G. Ordaz, and S. Satyapal, "The DOE National Hydrogen Storage Project: Recent Progress," Proceedings of the On-Board Vehicular Hydrogen Storage, Materials Research Symposium, 885E, A05-1.1, 2006.

103. S. A. Sherif, F. Barbir, and T. N. Veziroglu, "Principles of Hydrogen Energy Production, Storage and Utilization," *Journal of Scientific Industrial Research*, **62**, 46–63, (January–February 2003).

104. S. A. Sherif, F. Barbir, T. N. Veziroglu, M. Mahishi, and S. S. Srinivasan, *Hydrogen Energy Technologies* in Y. Goswami, (ed.) CRC Handbook of Hydrogen Energy Technologies, Boca Raton, Florida, 2006.

105. S. Turn, C. Kinoshita, Z. Zhang, D. Ishimura, and J. Zhou, "An Experimental Investigation of Hydrogen Production from Biomass Gasification," *International Journal of Hydrogen Energy*, **23**(8), 641–648 (1998).

106. U.S. DOE (2006), Hydrogen Program, http://www.hydrogen.energy.gov.

107. U.S. DOE Hydrogen Program Website (2006), http://www.hydrogen.energy.gov.

108. R.M. Zweig, "Pollution Solution/Revisited," *International Journal of Hydrogen Energy*, **17**(3), 219–225 (March 1992).

CHAPTER 8

CLEAN POWER GENERATION FROM COAL

James W. Butler and Prabir Basu
Department of Mechanical Engineering
Dalhousie University
Halifax, Nova Scotia
Canada

1	**INTRODUCTION**	**207**	
	1.1 Coal	208	
	1.2 Potential Pollutants from Coal	209	
	1.3 Motivation for Cleaner Energy from Coal	213	
	1.4 Cleaner Energy from Coal	214	
2	**PRECONVERSION**	**215**	
	2.1 Physical Cleaning	216	
	2.2 Chemical Cleaning	218	
	2.3 Biological Cleaning	219	
3	**COAL CONVERSION AND IN-SITU POLLUTION CONTROL**	**219**	
	3.1 Pulverized Coal Combustion	221	
	3.2 Fluidized Bed Combustion	224	
	3.3 Supercritical Boilers	230	

	3.4 Cyclone Combustion	231	
	3.5 Magnetohydrodynamics	232	
	3.6 Gasification	232	
4	**POSTCONVERSION CLEAN-UP**	**239**	
	4.1 Particulates	239	
	4.2 Gaseous Emissions	243	
	4.3 Heavy Metals	249	
	4.4 Solid Waste	250	
5	**CARBON DIOXIDE**	**251**	
	5.1 CO_2 Capture	252	
	5.2 Transportation	257	
	5.3 Sequestration and Utilization	257	
	5.4 Cost Implications	259	
6	**CONCLUSION**	**261**	

1 INTRODUCTION

Coal has consistently accounted for about 40 percent of the world's total electricity generating capacity since the early 1970s (Figure 1), despite the steady growth of the total generation capacity.[1] There is increased concern over global warming and emissions of sulphur dioxide, nitrogen oxides, and mercury from coal, but the low cost of electricity from coal has made it a necessary evil for many industries competing in the global marketplace. The world surely deserves pollution-free cleaner energy, at a cost that society can bear. Oil and gas provide a cleaner and more efficient means to generate electricity, but the cost is much higher. Geopolitical and economic factors make the price of oil and gas

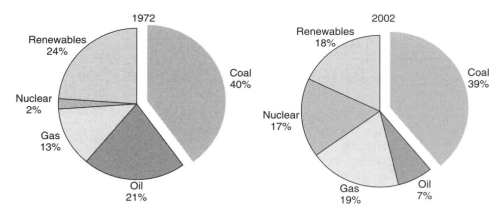

Figure 1 Worldwide energy production past and present (From Ref. 1, Key World Energy Statistics 2006 © OECD/IEA, 2006, p. 24.)

high and volatile, forcing utilities to rely on coal until other cleaner and more sustainable energy options mature and are available at competitive prices for the consumer.

Oil and gas prices show no signs of coming down significantly, public opposition to nuclear still remains strong and renewable energies still a low capacity option; electricity from coal is expected to increase to meet the rapidly increasing world energy demands. Low-cost electricity is crucial for the survival of some industries and is vital to the household expenditures of average people. So, instead of concentrating on the environmental ills of coal, it is worthwhile examining how clean we can make coal use while keeping the electricity price affordable.

The term *clean coal technology* has drawn criticism from environmental groups as the use of coal can never be entirely clean. A more appropriate term would be *cleaner coal technology*. The ultimate goal of *cleaner coal technology* is to produce coal-fired electricity generating plants with near-zero harmful emissions. This is the goal of the U.S. Department of Energy's FutureGen project[2] and the Near Zero Emission fossil fuel power plant project of the European Union.[3] Table 1 shows how the United States plans to meet the goal of zero emissions.[4]

1.1   Coal

Coal is formed from plant remains that have been compacted, hardened, chemically altered, and metamorphosed underground by heat and pressure over millions of years. It generally originates from swamp ecosystems. When plants die in a low-oxygen swamp environment, instead of decaying by bacteria and oxidation, their organic matter is preserved. Due to tectonic events, this organic material is buried by sedimentary loadings. Over time, heat and pressure remove the water and transform the matter into coal. Depending on the geological age, the organic

Table 1 Roadmap of Pollution Reduction from Coal-fired Power Plants Referred to Best Available Technology (Ref. 4)

		Reference (2003)	2010	2020
Pollution Mitigation	SO_2, reduction	98%	99%	>99%
	NO_x, (lb/10^6 BTU)	0.15	0.005	0.002
	Mercury capture	0%	90%	95%
	Byproduct utilization	30%	50%	100%
Plant efficiency (HHV)	Based on HHV	40%	45–50%	50–60%
Capital cost, $/kW	In 2003 dollar	1000–1300	900–1000	800–900
Electricity cost c/kWh	Based on coal $1.2/mBTU	3.5	3.0–3.2	<3.0

Source: Courtesy of U.S. Department of Energy/NTL Electric Power Research Institute, and the Coal Utilization Research Council.

material will transform into the following members of the coal family:

1. *Lignite*. This contains considerable amount of water and volatiles and some mineral matters. It is youngest in geological age.
2. *Bituminous*. Moderate amount of moisture, volatiles, and inorganic materials are in bituminous coal.
3. *Anthracite*. This contains very little water or moisture but a large amount of carbon. It is the oldest in geological age.

Figure 2 shows the full spectrum of these plant-based fuels, arranged in order of their geological age. It shows progressive changes in oxygen, volatile matter, and fixed carbon from wood to anthracite.

Nitrogen and sulphur appear in coal as organically bound pollutants that are oxidized during combustion to form harmful airborne gases such as nitric oxides (NO_x) and sulphur dioxide (SO_2).

Other impurities also become mixed with the coal during its formation:

- Ash—a quartz-based mineral and/or shale
- Pyretic sulphur (Fe_2S)
- Heavy metals either in elemental form or in the form of heavy metal ores, such as cinnabar (HgS), which releases mercury and sulphur upon heating

1.2 Potential Pollutants from Coal

Coal conversion generates a number of gaseous and solid pollutants:

1. Nitrogen oxides
2. Sulphur oxides
3. Carbon dioxide
4. Fine particulates
5. Heavy metals

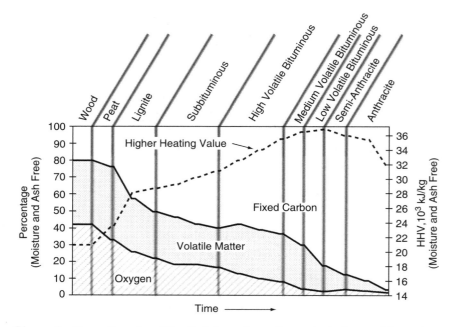

Figure 2   Properties of coallike fuel depends on the geological age of formation.

The following sections briefly describe the generation and effect of these pollutants.

Nitrogen Oxides (NO_x)

When exposed to high temperatures, nitrogen in the air and the fuel could oxidize, forming a number of different compounds, such as NO, NO_2, and N_2O. Nitrogen oxides (NO_x), which represents nitric oxide (NO) and nitrogen dioxide (NO_2), are responsible for the formation of acid rain, photochemical smog, and ground-level ozone. It could also cause adverse health effects for those with respiratory problems. Nitrous oxide (N_2O) is a greenhouse gas and also causes depletion of stratospheric ozone. The formation of NO and NO_2 is favored at elevated temperatures ($\sim 1{,}200°C$) as found in pulverized coal (PC) flames, while N_2O is favored at lower temperatures ($\sim 800°C$) like those found in fluidized bed combustors. The nitrogen in fuel is oxidized in all combustion temperatures, but that in air is oxidized generally above 1000°C.

Coal-fired plants are responsible for only a small portion of the anthropogenic NO_x emissions with the bulk coming from automobile emissions (Figure 3).[5]

Sulphur Dioxide (SO_2)

Sulphur dioxide (SO_2) is formed when the sulphur in coal, either in pyretic or organically bonded form, is oxidized. SO_2 is the leading contributor to acid rain

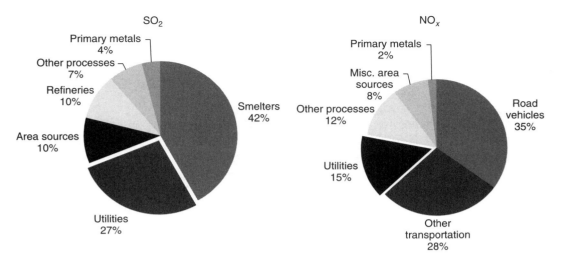

Figure 3 Man-made sources of NO$_x$ and SO$_2$ in Ontario, Canada. (Adapted from Ref. 5.)

formation and is an irritant to the lungs. Coal combustion makes up the majority of SO$_2$ emissions from utilities, which in turn account for about a third of the total SO$_2$ emissions from all sources (Figure 3).

Particulates

In a coal-burning plant a portion of the ash is released to the atmosphere through the stack, with the very fine particulates having harmful health effects. Particulate matter smaller than about 10 μm can settle in the bronchial tubes and lungs; particles smaller than 2.5 μm can penetrate directly into the lung; and particles smaller than 1 μm can penetrate into the alveolar region of the lung and tend to be the most hazardous when inhaled. Fine particulates can have significant harmful health effects if exposure occurs over extended periods.

Heavy Metals

Most heavy metals like lead, tin, and magnesium are collected in particulate control systems, but due to its low vaporization point (356°C), much of the mercury found in coal escapes into the environment, making coal combustion the primary source for mercury emissions. The mercury concentration in coal is high compared to that in the rock surrounding it, because vegetation growing in very wet conditions absorbs large amounts of mercury and coal deposits are formed from such wet decaying vegetation (peat). Mercury vapor in the flue gas precipitates out into the environment and bioaccumulates in organisms such as fish, where it is often transformed into methylmercury, a highly toxic organic compound. Fish species that are high up on the food chain contain high concentrations of mercury, because they eat many smaller fish that have small amounts of mercury in them. The U.S. Food and Drug administration has an action level for methyl mercury in commercial marine and freshwater fish that is 1.0 parts per million (ppm),[6] and in Canada the limit for the total of mercury content is 0.5 ppm.[7]

Carbon Dioxide (CO₂)

Carbon dioxide, a major greenhouse gas, is produced in all combustion processes involving fossil fuels as well in other industrial processes such as cement production and sweetening of natural gas. The absorption bands of CO_2 have a considerable overlap with the long wave infrared region of radiation (Figure 4). Thus, CO_2 allows the shorter wave radiation from the Sun to pass through, but traps the longer wavelength infrared radiation reflected from Earth's surface. This gives rise to the greenhouse-like warming of Earth, making CO_2 a greenhouse gas. Carbon dioxide stays in the atmosphere for hundreds of years, becoming a major threat to the biosphere, while water vapor, also having absorption bands in the infrared region, stays only for a few hours in the atmosphere before it is condensed. Thus, it is not considered a greenhouse gas.

The carbon content of coal is very high (50 to 89 percent) so it produces much larger amounts of CO_2 (carbon intensity) than produced by other fuels. Table 2 lists the amount of CO_2 produced per unit of energy released by different fossil fuels, with anthracite having the highest emission factor and natural gas having the lowest. Coal-fired power plants burning millions of tonnes of coal are therefore considered a major source of greenhouse gas. Since 1958, CO_2 in the atmosphere has seen a concentration increase of about 17 percent per year.

Despite divergent views on global warming, the scientific community largely accepts the temperature model that shows a large increase in average global

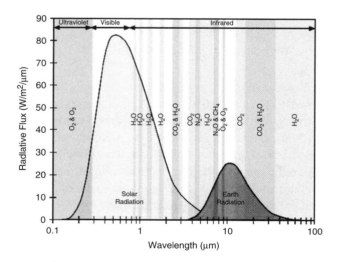

Figure 4   Atmospheric absorption spectrum.

Table 2 Emission Factors or Carbon Intensity (CO_2 Produced per Unit Amount of Heat Released) for Some Fuels

Fuel	Higher Heating Value, (MJ/kg)	Emission Factor (g_{CO2}/MJ)
Anthracite coal	26.2	96.8
Bituminous coal	27.8	87.3
Sub-bituminous coal	19.9	90.3
Lignite	14.9	91.6
Wood (dry)	20.0	78.4
Distillate fuel oil (#1)	45.97	68.6
Residual fuel oil (#6)	42.33	73.9
Kerosene	37.62	67.8
Natural gas	37.30 MJ/m^3	50

temperature over the past century compared to that of the previous millennium. Either way, it would be prudent to reduce anthropogenic CO_2 emissions rather than wait for more conclusive proof before irreversible damage has been done.

1.3 Motivation for Cleaner Energy from Coal

In December 1952, London became encased in a thick fog caused by a combination of particulate and gaseous emissions from coal-fired power plants and coal furnaces in homes; an unusually cold month; and an unfortunate climatic condition that caused the air in London to become stagnant. The fog caused the deaths of 4,000 people and led to Great Britain's Clean Air Act of 1956, resulting in a huge reduction in particulate emissions from coal-fired power plants. In 1963, the United States passed its own Clean Air Act.

As acid rain became an increasing problem in the eastern states and provinces, causing lakes and rivers to become barren and destroying forests, the United States signed the Convention on Long-Range Transboundary Air Pollution to help curb acid rain production. These regulations have helped to reduce the emissions of acid rain, causing pollutants from coal-fired plants (Figure 5).[8] The major efforts came in the 1970s and 1980s with the creation of international agreements such as Europe's Long-Range Transboundary Air Pollution Treaty, signed by 32 countries in 1979, that called for limits on emissions of harmful airborne pollutants.

In recent years, concern has grown about the levels of mercury found in the environment, even in places that are thousands of miles from any coal-fired power plant. This concern has caused the U.S. EPA (Environmental Protection Agency) to introduce the first ever mercury emissions standards through the Clean Air Mercury Rule, which requires reduction of mercury emissions by 70 percent (from 1999 levels) from coal-burning plants by 2015.

Government-regulated emission standards have been largely reactionary, as they are motivated by public awareness brought about by the twentieth century

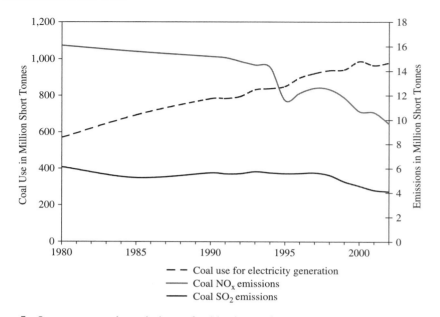

Figure 5   Improvements in emissions of acid rain causing pollutants from coal-fired power plants. (Adapted from Ref. 8.)

environmental movement. The driving force behind all research and development of cleaner coal technologies has likewise been government regulations. As companies are driven by financial decisions, government regulations have turned the environment into a commodity and introduced a financial disincentive for polluting instead of traditional punitive measures for polluting.

Most scientists now agree that the unprecedented levels of carbon dioxide concentration in the atmosphere is contributing to global warming, though by how much and what the consequences will be is still up for debate. Climate change affects the entire world, and as such, the Kyoto Protocol to the United Nations Framework Convention on Climate Change was created in 1997 to reduce the amount of carbon dioxide and other greenhouse gases entering the atmosphere. Currently, 163 states have signed the agreement and set individual goals for reduction of carbon dioxide. Incentive-based carbon dioxide reduction schemes may encourage power generation companies to greatly reduce their CO_2 emissions or turn to its sequestration.

1.4   Cleaner Energy from Coal

A number of technologies are available or under development to make the process of converting coal into a transmittable form of energy a less polluting process. These technical options can be broadly divided into the following three categories, based on which stage of the conversion process the pollutant reduction takes place:

1. Preconversion technology
2. In-situ control technology
3. Postconversion technologies

Preconversion technologies look to reduce the impurities from the raw coal before they are released in conversion and include physical, chemical, and biological cleaning. *In-situ technologies* help reduce the amount of noxious gases released from the conversion process. Some of these technologies are low NO_x burners (LNBs), fluidized bed combustion (FBC), supercritical boilers, gasification, and fuel cells. *Postconversion technologies* try to strip the flue gas of the harmful gases not eliminated through preconversion and conversion techniques. Some postconversion technologies are bag filters, electrostatic precipitators (ESPs), flue gas desulphurization (FGD), selective or selective noncatalytic reduction (SCR and SNCR) and carbon dioxide control involving CO_2 extraction and sequestration.

The purpose of this chapter is to present and explain the currently available cleaner coal technologies.

2   PRECONVERSION

An existing coal-fired power plant can improve its emissions to some extent without expensive modifications by using a cleaner variety of coal—that is, coal with a lower sulphur, ash, and heavy metals content. An added benefit of using a coal with low ash content is the savings on transportation cost of the raw coal to the plant and the ash to the disposal site. Cleaning the coal could potentially improve its utilization efficiency by up to 5 percent.[9]

High-quality coals are not as abundant as they were when electricity generation from coal first began. In addition, they are in high demand from the smelting industry. The end result is that using a cleaner raw coal is becoming less and less economical. This has lead to a growth in coal cleaning or benefaction, a process that makes possible the removal of mineral impurities that would otherwise be released in the conversion of the coal, polluting the atmosphere.

Increased mechanization in coal mining and a shift from underground to open-pit mining resulted in a higher ash content in the coal, making coal cleaning essential. As well, increasingly stringent sulphur and mercury emissions standards have made the removal of these substances a priority in the cleaning process. Coal cleaning mainly focuses on reducing sulphur, ash, and heavy metal in the coal. It can be classified into three main types:

1. Physical (most widely used)
2. Chemical
3. Biological

2.1   Physical Cleaning

Gravity Separation

Gravity separation technologies rely on the differences in specific gravity between the coal and the impurities it contains. The specific gravity of bituminous coal is in the range of 1.12 to 1.35, whereas that for pyrite is between 4.8 and 5.2, and for ash it is around 2.3. Thus, denser ash and pyretic sulphur can be separated from the lighter coal by static or dynamic means. Gravity separation most often involves wet separation, although it can also be accomplished through a dry process, thus reducing the energy required to later dry the cleaned coal, a necessity for the pneumatic transportation of coal in PC-fired plants. Gravity separation is used to clean coal particles larger than 0.5 inch (12 mm), as smaller pyretic particles would not settle out. Four of the more widely used gravity separation equipment are as follows:

1. Rotating drums
2. Concentrating tables
3. Cyclones
4. Dense-media vessels

In rotating drums, water and raw coal travel down an incline, where less dense coal remains near the surface while the more dense ash and pyrite particles settle to the bottom (Figure 6). At the exit, the top portion of the stream, containing mostly coal, is skimmed off and the rest is discarded.

Concentrating tables use flowing water to settle out the discard and carry the lighter coal particles across. The discard is periodically removed using moving riffles.

Cyclones using either water or a dense media to separate the coal from the discard using the centrifugal forces from the tangential inlet of the cyclone. The lighter coal particles escape through the top of the cyclone, while as the heavier

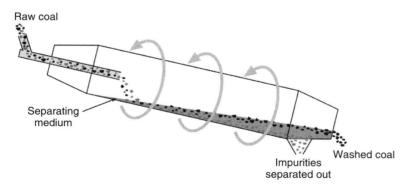

Figure 6   Rotating drum coal washing.

discard is held to the walls by the tangential forces and spirals down and out the bottom.

In dense-media vessels, the raw coal flows through a large vessel with a liquid that has a specific gravity just higher than that of the coal (usually 1.45 to 1.65). The commonly used medium is water with finely ground magnetite suspended in it. After the mixture has passed through the vessel and the denser pyrite and ash have settled out, the magnetite is separated from the coal–water mixture using magnetic separators and is then recycled.

Agglomeration

Agglomeration is used to clean coal fines ($< 100 \, \mu$m) and can recover very small coal particles $< 10 \, \mu$m. The naturally hydrophobic coal fines are suspended in an aqueous solution, a light oil is added, and the solution is agitated. The oil preferentially wets the coal, and the agitation causes the fines to agglomerate into larger particles, 1–2 mm in size. These larger particles are then screened from the fine impurities.

The coal particles can be further agglomerated by pelletization with a binder such as asphalt. Agglomeration is becoming increasingly costly because it uses a large amount of expensive light oil to agglomerate the coal, which cannot be reused.

Froth Flotation

Froth flotation involves passing air up through an aqueous solution containing a frothing agent and pulverized coal (Figure 7). The hydrophobic coal particles attach to the air bubbles, rise to the top, and are skimmed off with the froth, while the mainly hydrophilic impurities sink to the bottom. In some flotation processes, a modifier (i.e., fuel oil) is needed to increase the hydrophobicity of the coal. Froth flotation has three problems:

1. Entrainment of small ash particles by bubbles into the froth phase
2. Low probability of collision between small coal particles and air bubbles
3. Pyrites that have a natural hydrophobicity, causing them to attach to the bubbles

Column flotation increases the likelihood that small coal particles will come in contact with air bubbles, lifting them to the surface, and that entrained ash particles will be dropped out (Figure 7). Also, new micro-bubbling techniques can increase the coal particle recovery by increasing the odds that ultra-fine coal particles will come into contact with air bubbles.

High Gradient Magnetic Separation (HGMS)

This process relies on the magnetic properties of pyrite (FeS_2) to separate it from the coal. A coal/liquid slurry is passed through high-intensity magnetic fields,

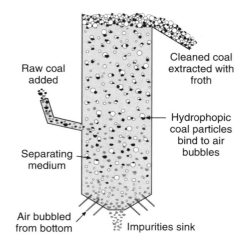

Raw coal added

Cleaned coal extracted with froth

Hydrophopic coal particles bind to air bubbles

Separating medium

Air bubbled from bottom

Impurities sink

Figure 7 Column froth flotation coal cleaning.

where the magnetic impurities are drawn to the sides and discarded. The particles in the slurry must have a size distribution such that 70 percent are smaller than 76 microns to maximize the capture of mineral matter in the coal.[10] The magnetic properties of the impurities can be enhanced by chemical fragmentation of the coal before going through the HGMS.[11] Problems with this process include the large capital investment required and its inability to remove the majority of ash particles that are nonmagnetic.

2.2 Chemical Cleaning

Cleaning of coal through chemical means is mainly concerned with removing the organically and inorganically (pyrite) bound sulphur from the coal prior to conversion. Some of the methods of chemically removing the sulphur from the coal follow:

- Chlorination
- Direct oxidization
- Indirect oxidation
- Mild hydrogenation

The operating conditions of chemical cleaning processes are very severe, with high temperature, long retention time, and high alkalinity leading to high-cost equipment and hazardous working conditions. On the one hand, chemical cleaning does have the benefit of removing organically bound sulphur, something physical cleaning cannot accomplish. On the other hand, chemical cleaning alone is unlikely to meet the sulphur removal standards required, so it is rarely used in commercial plants.

2.3 Biological Cleaning

A number of microorganisms (mainly bacterial species) can eliminate the pyretic and/or organically bound sulphur. These bacteria fall into three categories, depending on the type of sulphur they can remove:

1. *Obligate autotrophs* —oxidize pyretic sulphur only
2. *Facultative autotrophs* —oxidize pyretic sulphur and some organically bound sulphur compounds
3. *Heterotrophs* —oxidize organic compounds only.

Research has focused on the obligate autotrophs, as the majority of the sulphur found in coal is in the pyretic form. The reaction rates for sulphur removal are still too slow for commercial applications, so instead of using the bacteria to dissolve the pyrite, the hydrophilic bacteria is used to modify the surface chemistry of the pyrite in order to enhance the physical cleaning processes. Biological cleaning is still at a bench scale stage, and requires further research to assess its commercial viability for removing sulphur from coal.

3 COAL CONVERSION AND IN-SITU POLLUTION CONTROL

As emission standards become more stringent, new and cleaner ways of converting coal to usable forms of energy are being developed and implemented (see Table 3).[12] Energy from coal follows one of two conversion routes: the combustion of coal to produce heat energy, which is used to drive a steam turbine, or the gasification of coal to produce a combustible gas (syngas) that can be used to generate heat, electricity, or hydrogen gas for sale.

Pulverized coal (PC) combustion, introduced in 1910, dominated as the most advanced coal-fired generating technology until fluidized bed combustion arrived on the scene in the 1980s. Advanced steam turbines with higher steam temperature greatly improved the overall power generation efficiency of such plants, raising it

Table 3 Improvements in the Emissions from Coal-fired Power Plant (Adapted from Ref. 12)

Plant Technology	Period of Use	Efficiency (HHV)	SO_2 (g/kWh)	NO_x (g/kWh)	CO_2 (g/kWh)	Ash + Sorbent (g/kWh)	Waste Heat in Cooling Water, MJ/kWh
PC	1950–70	30%	0.029	0.0034	1080	45 (ash only)	5.6
PC + FGD + SCR	Present	41%	0.0104	0.00029	770	84	4.0
PFBC combined cycle	Present	39%	0.0059	0.00058	815	99	3.6
IGCC	Present	42%	0.00015	0.00029	745	40.7	3.2
Natural gas-fired combined cycle	Present without SCR	52%	0	0.00031			2.3

from 15 percent to 39 percent.* Combined cycle power plants, which use both steam and gas turbines, could increase this efficiency to 48 percent using an integrated gasification combined cycle (IGCC)[12] and has the potential of raising it above 50 percent with a partial gasification system.[13] Such higher-efficiency plants cause less thermal as well as gaseous pollutants per unit energy produced (Table 3). As shown in Table 1, efforts toward zero emission power plants strive to reduce the emission and rejections to a near zero level using the gasification route.

To meet the environmental regulations of governments and the economic need of the consumers, advanced power generation technologies strive to achieve two things:

1. Reduce the emission of harmful gases per kWh generated
2. Reduce the cost per kWh generated

These objectives are met either by increasing the energy conversion efficiency or by a conversion process with less inherent generation of pollutants. Some of the advanced technologies available for generation of cleaner energy conversion from coal are as follows:

- Rankine cycle plants operating on a supercritical steam cycle
- Integrated gasification or partial gasification combined cycle plants
- Pressurized fluidized bed combustion plants
- Fuel cells

Coal combustion is the high-temperature oxidation of carbon and hydrocarbon content of the coal. The basic equation is as follows:

$$C + O_2 = CO_2 + \text{Heat} \tag{1}$$

$$C_n H_m + O_2 = H_2O + CO_2 + \text{Heat} \tag{2}$$

These equations are the basis of mass balance or stoichiometric calculation needed to calculate the amount of air required and the amount of product produced, but these alone do not give the complete picture of the actual combustion process.

The combustion process requires transportation of the necessary amount of oxygen to the fuel surface, removal of product gases, a favorable temperature for reactions to occur, and a sufficient time for the reaction to complete. These three requirements led to the famous three T requirements for combustion:

1. *T*urbulence for efficient transport of oxidant and products
2. *T*emperature for necessary rate of reaction
3. *T*ime for completion of the reactions

The transport of oxygen from the air to the surface of fuel is a major factor governing the combustion process. Coal being solid does not mix as easily with

---

* All efficiencies are expressed in terms of higher heating values (HHV). Some efficiencies were converted from LHV using conversion factor of 0.8675, using HHV = 28000 kJ/kg and [H] = 5%.

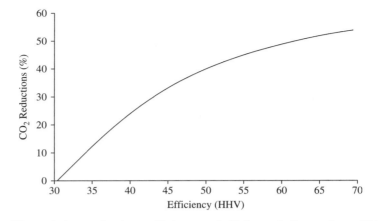

Figure 8 CO_2 emissions reductions with increased efficiency (referenced to a 30% efficient plant). (Adapted from Ref. 15).

the oxygen as gaseous or liquid fuels do. As a result, a good gas–solid contact is vital for coal combustion. The furnace design also has to allow enough time for the coal particles to complete their combustion. Detailed discussion of the combustion process, though an integral part of clean coal technology, is beyond the scope of this chapter. Authors may be directed to Chapter 4 in Basu.[14]

Although coal combustion has been in use for centuries, it is only in the last few decades that efforts have been made to lessen the harmful effects of this process. Research has gone into developing advanced, more-efficient plants that produce less emissions. Figure 8 shows how the CO_2 emissions are reduced when the plant efficiency is increased above its 30 percent level of the 1970s.[15]

In order to reduce the pollutants generated, six in-situ coal conversion systems are available:

1. Pulverized coal (PC) boiler using a low NO_x burner (LNB)
2. Fluidized bed combustion
3. Supercritical boiler technology
4. Cyclone combustion
5. Magnetohydrodynamics
6. Gasification

A brief description of these technology options is described in this section.

3.1 Pulverized Coal Combustion

In this type of combustion, finely ground (pulverized) coal is burnt in a furnace to generate steam that either expands in a steam turbine to generate electricity

(Carnot cycle) or provides process heat. Pulverized coal combustion (PC) started in the 1910 and is still the workhorse of coal-fired power plants around the world.

Energy conversion efficiency depends on the Carnot cycle efficiency, steam generation efficiency that includes combustion efficiency of coal, and the turbine efficiency. In the simplest form, the Carnot efficiency is written in terms of the highest (T_{max}) and the lowest (T_{min}) cycle temperatures of the working fluid. For Rankine cycles steam plants these are the temperatures of the steam at the inlet and exhaust of the turbine respectively. In gas/steam turbine combined cycles, T_{max} is the gas temperature at the inlet of the gas turbine (shown later in Figure 13).

$$\eta_{carnot} = 1 - \frac{T_{min}}{T_{max}} \tag{3}$$

The condenser exhaust temperature (T_{min}) is dependent on the condenser pressure and cooling water temperature, while the turbine inlet temperature (T_{max}) is dependent on the maximum steam temperature the boiler can deliver.

The efficiency of early (1900s) plants was low, in the range of 15 percent and steam temperature and pressure were modest 180°C/1.0 MPa. The efficiency rose to about 29 percent, steam temperature as pressure rose to 538°C/1.4 MPa and reheating/feedwater heating was introduced. The plants were still based on a very basic boiler arrangement without environment control systems.[16]

Up to the 1960s the primary design changes that led to increases in thermal efficiency included increasing the steam pressure, the number of reheat cycles, and the amount of feedwater heating using multiple extraction points.[17] Subcritical pressure cycles (16.6 MPa and 538°C) with a single-stage reheat and feedwater heating pushed the efficiency to about 34.5 percent. The advent of supercritical boilers pushed the frontier further to 38.9 percent with 27.5 MPa, 565°C steam. With further rise in steam pressures and temperatures (31.1 MPa, 604°C) and double reheat, ultra-supercritical plants are poised to raise the plant efficiency above 41 percent.[18]

Huge advances in control systems for coal-fired plants have also contributed to the increases in efficiency since the advent of the microprocessor in the 1970s.

The technology behind conventional pulverized coal-fired combustors is well developed, inexpensive, and capable of very large generating capacities. These types of combustors are likely to remain the dominant type of coal-fired furnace in the world for years to come. But, because of the unacceptable environmental performance of its basic form, new technologies have been developed to reduce the harmful emissions from PC boilers.

Pulverized coal-fired furnaces need high flame temperatures ($\sim 1500°C$) for rapid combustion of very fine (90% $<76\,\mu m$) coal particles.[19] This rapid combustion is needed to prevent coal fines escaping the furnace unburnt, but its high flame temperatures cause the oxidation of nitrogen in the combustion air and in the fuel into NO_x. Furthermore, the high flame temperature does not allow for in-situ sulphur capture through injection of limestone into the furnace.

Four means of mitigating the problem of NO_x formation in a PC furnace are as follows:

1. Using low excess air (low amount of air above the stoichiometric requirement)
2. Reducing the peak flame temperature
3. Firing a mixture of O_2 and recycled flue gas
4. Using low NO_x burners

Low NO_x Burners (LNB)

Low NO_x burners are the most cost-effective method of reducing NO_x emissions from an existing PC-fired power plant, but it can reduce NO_x emissions only up to 50 percent. In a typical low NO_x burner (Figure 9) the NO_x formation is reduced by two factors:

1. Staging the combustion air so that an inner fuel-rich zone is created, causing the fuel nitrogen to be released as N_2
2. Decreasing the temperature of the outer fuel-lean zone reducing the formation of thermal NO_x

The reduction of the flame temperature is difficult to achieve without sacrificing the combustion efficiency of char particles. Low excess air could reduce NO_x formation and is used in some boilers, but it also affects the combustion efficiency.

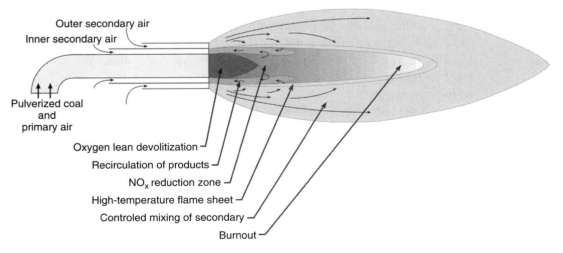

Figure 9 Low NO_x burner.

NO Reburning

For further NO_x reduction in the furnaces of PC boilers, a secondary fuel (CH_4, fuel oil, or high-volatile coal) can be burnt downstream of the main combustion zone to reduce the NO_x already formed by converting it to N_2. The NO_x is reduced by the combustion of hydrocarbon radicals in an oxygen deficient environment.

$$2NO + 2C_nH_m + (2n + m/2 - 2)O_2 \rightarrow N_2 + 2nCO_2 + mH_2O \qquad (4)$$

Char can also react to reduce the NO_x. The problem with reburning is that its oxygen deficient environment reduces the combustion efficiency and increases unburnt hydrocarbons or char.

3.2 Fluidized Bed Combustion

Fluidized bed combustion is the second revolution in the art of coal combustion after PC combustion. In a fluidized bed, air is passed through a grate supporting a mass of inert solids at a velocity such that the solid mass behaves as a fluid, giving it the name *fluidized bed*. In fluidized bed combustion, crushed coal burns in a suspension (bed) of highly agitated, hot, inert solids. Fuels constitute only a small fraction (typically 1 to 3 percent) of the total solids in this bed. The intense mixing, relatively uniform temperature, and large thermal inertia of the combustion zone allow fluidized bed combustion to burn most types of fuels, good or bad. This flexibility allows the generation of power from many inexpensive low-grade fuels such as petcoke, waste coal, and lignite, which cannot be burnt efficiently in pulverized coal-fired boilers.

Besides fuel flexibility, fluidized bed combustion offers reduced NO_x emission (due to the near absence of thermal NO_x formation) and the capability of sulphur capture in the furnace, 90 percent or more. Thus, a fluidized bed boiler can meet most of today's environmental standards without the postcombustion clean-up systems required by a PC boiler. These features are a direct result of its low (800°C–900°C) combustion temperature. However, this low temperature leads to increased emissions of N_2O, a greenhouse gas.

Thousands of subcritical fluidized bed boilers with capacities from a few hundred kW up to 350 MW are in operation worldwide, with a 485 MWe supercritical boiler to begin operation in Lagisza, Poland.[19]

Fluidized bed boilers are of two principal types:

1. Bubbling fluidized bed (BFB)
2. Circulating fluidized bed (CFB)

Either type can be designed to operate at an elevated pressure, making it a pressurized fluidized bed combustor (PFBC) that can run a combined cycle plant.

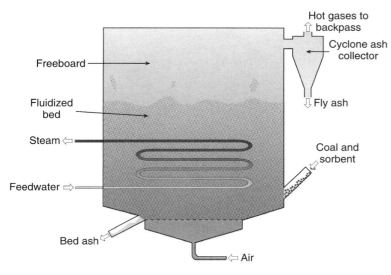

Figure 10   Bubbling fluidized bed combustor.

Bubbling Fluidized Bed

The mean gas velocity through a bubbling fluidized bed combustor is typically in the range of 1.0 to 2.5 m/s. Coal particles with diameters of upwards of 6 mm are fed into a fluidized bed and heated by the hot inert solids making up the bed. The heat energy produced from the combustion of the coal is absorbed by the inert solids and then transferred to tubes exposed to the bed (Figure 10). Flue gas leaves the bed at about 800° to 900°C and passes through the relatively empty space above it (freeboard) before leaving the furnace to enter the back-pass of convective section of the boiler.

Some of the solids and fine char particles escape the bed and furnace as fly ash, but the majority of the ash is drained from the bottom of the bed (80 percent, compared to PC at 20 percent), reducing the amount of particulates exiting the furnace. BFB boilers without heat-absorption tubes in the bed are particularly suitable for biomass and waste firing because of their ability to sustain combustion of fuels of low calorific value.

Sulphur capture in the BFB furnace can be accomplished by using a sorbent such as limestone ($CaCO_3$) as the bed material. Limestone captures the sulphur dioxide generated from coal combustion, through the following reactions:

$$CaCO_3 \rightarrow CaO + CO_2 \tag{5}$$

$$CaO + SO_2 + \tfrac{1}{2}O_2 \rightarrow CaSO_4 \tag{6}$$

These reactions progress best in the temperature range of 800° to 900°C, which exists in fluidized bed furnaces.

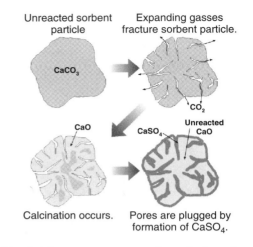

Unreacted sorbent particle

Expanding gasses fracture sorbent particle.

$CaCO_3$

CO_2

CaO

$CaSO_4$

Unreacted CaO

Calcination occurs.

Pores are plugged by formation of $CaSO_4$.

Figure 11 Sulphur capture process for a single sorbent particle.

Rarely is calcium oxide converted completely into calcium sulphate, due to the plugging of sorbent particle pores caused by the increased molar volume of the sulphated sorbent molecules as the sulphur dioxide is captured, making most inner surface areas of the sorbent particle unavailable for sulphur capture (Figure 11). Fine particles could increase the effective surface area for sulphur capture, but the entrainment of fine particles prevents the use of sorbents any finer than 500 micron in a bubbling fluidized bed. Thus, to capture one mole of sulphur a bubbling bed may need as much as 2.5 to 3.5 moles of CaO, meaning that 2.5 to 3.5 times the stoichiometric amount of sorbent is required for effective capture of the sulphur. This calcium-to-sulphur molar (Ca/S) ratio is an important parameter in fluidized bed sulphur capture.

From the reaction in equations (5) and (6), it is seen that for every mole of sulphur dioxide captured, one mole of carbon dioxide is released, but due to the plugging of pores, for every mole of sulphur captured about 3 moles of extra CO_2 are produced in a BFB boiler. This extra CO_2 emission is also present in PC boilers with a flue gas desulphurization (FGD) scrubber, although to a lesser extent.

An atmospheric BFB combustor requires a large footprint, owing to its relatively low grate heat release rates of 1 to 2 MW/m^2.[20] As well, in a BFB furnace, a large number of feed points are required. For example, a 160 MWe unit requires 120 feed points.[20] These two factors limit the capacity of BFB boilers to small industrial applications, as well as for combustion of waste materials.

Circulating Fluidized Bed Boiler
The other member of the fluidized bed family is the circulating fluidized bed (CFB), which uses much higher gas velocities (4–6 m/s), compared to BFBs in

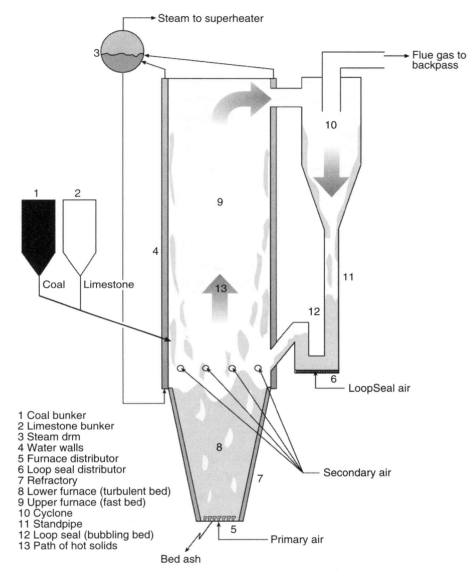

Figure 12 Circulating fluidized bed combustor.

the furnace, causing the bed of solids to be entrained out of the furnace. The solids are then collected by a cyclone or other type of separator, and recycled back at a sufficiently high rate by means of a solid recycle valve such as a loop seal (Figure 12). This creates a special hydrodynamic condition known as a *fast bed* in the furnace, characterized by intense internal as well as external recirculation of solids.

The combustion air is fed into the furnace in stages. Thus, the lower section of the furnace, which operates in a substoichiometric condition, is refractory lined, while the enclosure of the upper furnace that operates with about 20 percent excess air is made of heat-absorbing evaporator tubes similar to that in PC boilers. Depending on the heat absorption requirements, additional tube panels in the furnace or an external bubbling bed heat exchanger may be needed to maintain the furnace in the desired temperature range (800°–900°C) for a fluidized bed combustor.

The internal and external circulation of solids allows for long residence times even for very fine ($\sim 10 \, \mu$m) coal and sorbent particles. Thus, a CFB boiler can use relatively fine (~ 200–$300 \, \mu$m) sorbent particles, increasing the effective surface area for capture and resulting in a reduced calcium to sulphur molar ratios of 1.5 to 2.0[14] and as a result it achieves reduced CO_2 emissions compared to that for BFB boilers.

The high residence time also leads to an increase in combustion efficiency (~ 99 percent). Other advantages of CFB combustors include a small footprint owing to its higher grate heat release rate (3–5 MW/m^2), fewer feed points (four points for a 190 MWe plant) and greater fuel flexibility.[14] Furthermore, the staging of combustion air to provide favorable conditions for reduction in NOx emission is possible in a CFB furnace.

FBC Repowering

One option for reducing the emissions from older PC-fired plants is to retrofit them for fluidized bed firing. This is an alternative to the addition of a scrubber and LNB to an old PC boiler. There are three main reasons for repowering an aging PC plant:

1. Deterioration of the fuel supply over time
2. Reducing the cost of expensive support fuel (oil/gas)
3. Reducing plant emissions to meet increasing standards

The volumetric and grate heat release rates of a PC boiler significantly overlap those of a CFB boiler.[13] As a result many old PC boilers can be adapted to CFB firing without major modifications to the plant, or a new CFB boiler can be built in the footprint of the existing PC boiler.[21]

The least invasive means of repowering would involve modification of the furnace to include a compact separator (cyclone or impact separator) and a solid recycle system.[22]

In the other repowering option, the entire boiler is replaced with a new CFB boiler. Such a procedure can extend the life of older plants by as much as 25 years. In addition, auxiliary fuel consumption can be substantially reduced.[23]

Repowering provides a PC plant with all the environmental and fuel flexibility benefits of a CFB plant for less cost than is needed to build a new CFB plant. These benefits include a reduction in NO$_x$ as well as in-situ sulphur capture potential. This may prove in some cases less expensive than adding a scrubber for environmental upgrade on an old PC plant.

Pressurized Fluidized Bed (PFBC)

In a PFBC the hot pressurized flue gas is filtered and enters the gas turbine at around 800°C for Brayton cycle power generation. The waste heat from the gas turbine and the combustion heat in the PFBC boiler are used to generate steam that drives a steam turbine in a traditional Rankine cycle. This combined cycle can attain efficiencies up to 40 percent with a subcritical boiler. The high efficiency is due to the higher turbine inlet temperature (T_{max}) of its working fluid (equation 3).

Figure 13 illustrates how the combination of Rankine and Brayton cycles reduces the relative losses to improve the cycle efficiency. It also illustrates how the addition of a reheater could improve plant efficiency.

The furnace of a pressurized fluidized bed boiler is operated at an elevated pressure such that the hot flue gas produced can be expanded in a gas turbine.

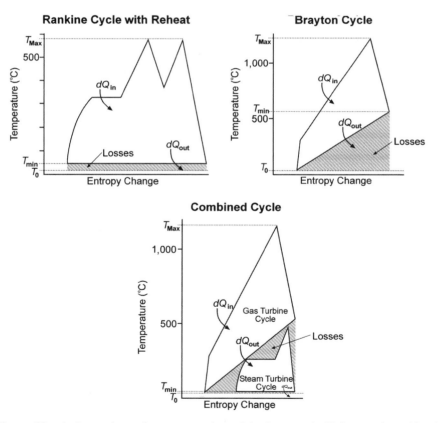

Figure 13 A thermodynamic representation of the improved efficiency of combined.

Such a high-pressure operation gives a grate heat release rate several times higher than that of an atmospheric FBC and increased combustion efficiency.

The hot flue gases from the PFBC must be stripped of both particulate matter and sulphur compounds to avoid erosion and corrosion, respectively, of the turbine blades. Currently, only six PCFB units are in operation around the world.[19] Availability of higher efficiency systems such as IGCC or PGCC, which still use a fluidized bed for the coal conversion, reduced interest in this technology.

3.3 Supercritical Boilers

Supercritical or ultra supercritical boiler technology can be applied to conventional PC-fired combustors as well as to fluidized bed combustors. Conventional boilers generate steam at around 16 MPa and 540°C, but supercritical boilers generate steam at much higher pressures (exceeding 22.1 MPa) and temperature (>374°C). Ultrasupercritical boilers operate with steam pressures and temperatures as high as 35 MPa and 760°C.[24] Owing to their high steam temperature and pressure supercritical boilers give plant efficiencies well exceeding 40 percent (HHV basis) compared to 35 percent for current subcritical units[25] leading to a reduction in emissions per kWh from the plant (Figure 14).[26] For example, a 300MWe supercritical boiler reduces carbon dioxide emissions by 137,000 tons/yr even with a very modest efficiency gain of 1.7 percent.[19]

Supercritical boilers use a once-through type of water/steam flow under forced circulation, eliminating the need to use expensive steam drums, downcomers, and so on. This is the main reason why supercritical boilers are within 3 to 5 percent of the cost of subcritical boilers, despite their need for high-pressure equipment and high-pressure pumps.

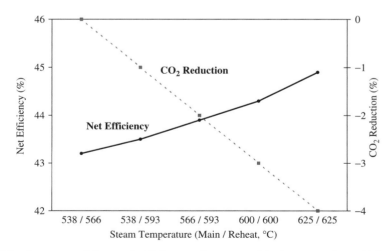

Figure 14 Steam condition and efficiency of a supercritical power plant. (Adapted from Ref. 2.)

PC supercritical boilers suffer from the problem of very high and uneven heat flux distribution around the furnace. Uneven heating of steam/water in the evaporator panels could cause evaporator tube failure if the heat flux to one section is too high compared to another. In a PC boiler, this problem is overcome by several means, such as the use of the following three means:

1. An expensive spiral tube arrangement where the tubes wrap around the boiler, so that each tube is exposed to the same sections of varying heat flux

2. Riffled tubes, which help to reduce the surface temperature of the tubes

3. A multipass system where all tubes go up one wall of the boiler, meet at a header at the top, go down another wall of the boiler.

Circulating fluidized bed boilers are relatively free from these problems as its lower (compared to PC boiler) heat flux and relatively uniform distribution around the furnace periphery, eliminate the need for complex tube construction.

CFB firing of a supercritical boiler is relatively new. In 2006, more than 500 PC supercritical plants were in operation around the world—most (46 percent) in the former Soviet Union.[27] Large increases in the number of additional units are expected with India and China choosing to adopt supercritical boilers as their standard for plants with generation above 800 MWe.[9]

Owing to their proven records, supercritical plants with postcombustion clean-up (for PC firing) and in-situ cleaning (for CFB firing) may be more attractive to the utility industries than IGCC plants for generation of clean, reliable, and cost-effective power from coal. Unlike IGCC such plants are suitable for both high- and low-rank fuels.

3.4 Cyclone Combustion

Cyclone combustion technology offers high heat generation per volume (18.5 GJ/hr·cm^3) and offers a good retrofit option for existing oil and gas fired plants as prices continue to increase.[11] It is useful for coals with a low ash melting temperature, which is difficult to burn in PC boilers. In a cyclone combustor, high-velocity air carrying coal particles is tangentially injected into a horizontal cyclone producing spiraling of the combustion gas around the furnace (Figure 15). Its uncooled furnace generates high temperature (1650°C) causing the ash to slag and allowing for very high combustion efficiencies (>99%).[11]

Staged combustion can help to prevent the production of excessively high levels of NO_x in the high temperature environment. The sulphur dioxide is removed only to a limited extent by injection of sorbents into the furnace along with the coal air stream. High temperatures also allow for removal of 70 percent of the ash, as liquid slag reducing the load on the particulate capture system.

In 2006, more than 100 cyclone combustion units were in use; however, owing to its high NO_x emission, no new units are planned or under construction in the United States.

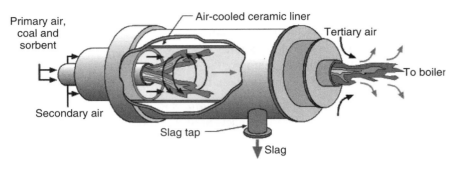

Figure 15 Cyclone combustor.

3.5 Magnetohydrodynamics

Magnetohydrodynamics (MHD) uses very hot ($\sim 2,500^\circ$C) combustion gases seeded with a compound that is easily ionized (K or Ce), passed at high velocities through a magnetic field to generate the flow of electrons. If combined with a Rankine cycle downstream of the MHD, efficiencies of 35 to 52 percent are expected.[28] MHD is still in the research stage with a number of hurdles to be overcome, such as the development of powerful superconductors, mitigation of thermal NO_x emissions, and the reduction of the high capital cost.

3.6 Gasification

Gasification is essentially the process of converting the organic material of coal into a gaseous form. This is accomplished by reacting coal in a gasifier, operated in a substoichiometric or oxygen deficient environment. It has a special place in the world of cleaner coal technology, as it is the centerpiece of near-zero-emission power plants and could provide valuable byproducts such as hydrogen from coal. The energy density of hydrogen is very high (121 MJ/kg), about five times that of raw coal; therefore, it is suitable for many other applications such as transportation and use in fuel cells.

The first stage of gasification, called *pyrolysis*, occurs from 400°C and up in the absence of oxygen. Here coal is converted into char and hydrogen-rich volatiles. In the second stage (700°C and up), air or oxygen is added to gasify the char, yielding a combustible gas and ash. The key reactions involved in gasification are

$$
\begin{array}{llrl}
\text{Combustion:} & C + O_2 \rightarrow CO_2 & +393 \text{ kJ/mol} & (7) \\
\text{Boudouard reaction:} & C + CO_2 \rightarrow 2CO & -172 \text{ kJ/mol} & (8) \\
\text{Carbon-steam reaction:} & C + H_2O \leftrightarrow CO + H_2 & -131 \text{ kJ/mol} & (9) \\
\text{Shift reaction:} & CO + H_2O \leftrightarrow CO_2 + H_2 & +42 \text{ kJ/mol} & (10)
\end{array}
$$

Energy for the endothermic gasification reactions, shown in equations (8) and (9) is provided by a certain amount of combustion, shown in equation (7), also taking place in the gasifier. Regulating the amount of oxygen fed into the gasification chamber controls this extent of combustion. The shift reaction shown in equation (10) is encouraged if the desired product of gasification is hydrogen.

The air—in which nitrogen remains in the product gas, reducing its heating value—provides the oxygen for the reactions in equations (7) to (10). If gasification uses pure oxygen instead of air, a *syngas* consisting of CO_2, CO, and H_2 is produced that has twice the heating value obtained by air-blown gasification and about 20 percent the heating value of natural gas. Oxygen-blown gasification has the added benefit of carbon dioxide capture, as its end product is a high-pressure stream of water and carbon dioxide. A simple condensing out of the water would leave a stream of CO_2 for sale or sequestration. The thermal efficiency of an oxygen-blown gasifier, including carbon dioxide capture and sequestration, is about 73 percent (IEA[29]). A combined cycle using such a gasifier would have an overall plant efficiency in the range of 43 to 52 percent.[28]

There are three generic types of gasification reactors:

1. Fixed bed
2. Fluidized bed
3. Entrained flow reactors

The main difference between these types lies in the gas–solid contacting process employed in the gasification chamber. Table 4 shows that the mode of contact has important influences on the gasifier design and performance.[30]

On one hand, the gas velocity in a fixed-bed gasifier is below that required for fluidization of the coal or ash particles in it. Bubbling and circulating fluidized bed gasifiers, on the other hand, use gas velocities above the minimum fluidizing and terminal velocities of the average coal or ash particles, respectively. Entrained flow gasifiers operate similar to the burners of a PC boiler. The gasification medium (oxygen, steam, or air) transports pulverized coal particles through the gasifier at a velocity well above the transport velocity of the coal particles. Since the particles are not recycled, their residence time in the reactor is low.

Fixed beds are less expensive but are suitable only for small-capacity units. Fluidized beds are more expensive and are suitable for large to medium size units. Entrained bed gasifiers are generally used for large-capacity units.

A gasification-based plant can produce energy in the form of steam, electricity, or syngas for production of chemicals or hydrogen. It is especially important for the hydrogen economy of the future. Figure 16 shows a flow chart of one energy conversion option for gasification.

Integrated Gasification Combined Cycle (IGCC)

An integrated gasification combined cycle (IGCC), as the name implies, would gasify coal into fuel gas to fire a gas turbine and use the waste heat to generate

Table 4 Important Characteristics of Generic Types of Gasifiers (Adapted from Ref. 30)

Gasifier Type	Moving Bed		Fluidized Bed		Entrained Flow
Ash conditions	Dry Ash	Slagging	Dry Ash	Agglomerating	Slagging
Fuel Characteristics:					
Fuel size limits	6–50 mm	6–50 mm	<6 mm	<6 mm	0.1 mm
Preferred feedstock	Lignite, reactive bituminous coal, anthracite, wastes	Bituminous coal, anthracite, petcoke, wastes	Lignite, reactive bituminous coal, anthracite, wastes	Lignite, bituminous, anthracite, cokes, biomass, wastes	Lignite, reactive bituminous coal, anthracite, petcokes
Ash content limitations	None	<25%	None	None	<25%
Preferred ash melting temp. (°C)	>425	<650	>925	>925	<1035
Operating Characteristics:					
Exit gas temperature (°C)	425–650[a]	425–650	925–1035	925–1035	>1260
Gasification pressure (MPa)	3.0+	3.0+	0.1	0.1–3.0	5.0
Oxidant requirement	Low	Low	Moderate	Moderate	High
Steam requirement	High	Low	Moderate	Moderate	Low
Unit capacities (MWth)	10–350	10–350	100–700	20–150	Up to 700
Key Distingushing Characteristics	Hydrocarbon liquids in raw gas		Large char recycle		High sensible heat energy in raw gas
Key Technical Issue	Utilization of fines & hydrocarbon liquids		Carbon conversion		Raw gas cooling

[a]Moving-bed gasifiers operating on low rank fuels have temperatures lower than 425°C

234

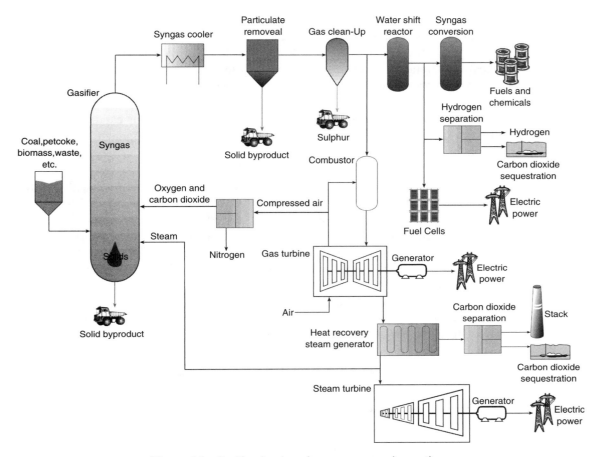

Figure 16 Gasification-based energy conversion options.

steam to run a steam turbine. As such, it enjoys the benefit of higher peak working fluid temperature ($T_{\max}$) with higher-efficiency power generation of the Brayton and Rankine combined cycle (see equation (3) and Figure 14).

In a typical IGCC plant (Figure 16), prepared coal is fed into a gasifer, which could be of fluidized or entrained type. The coal is gasified into CO or H_2. Due to the reducing condition in the gasification chamber, the majority of the nitrogen and sulphur in the coal is not oxidized, reducing the production of the atmospheric pollutants NO_x and SO_2. The majority of the nitrogen from the coal is released as N_2 and NH_3. The latter is removed from the syngas prior to combustion to avoid corrosion of the gas turbine.

Sulphur reacts with hydrogen in the gasification chamber to produce H_2S, which is again removed from the flue gas prior to combustion to prevent corrosion. This is accomplished using amine scrubbers that generate organic sulphur or sulphuric acid, both of which are saleable byproducts.

The particulates are removed using hot gas filters or a water scrubber that requires the temperature of the syngas to be greatly reduced. For sulphur removal the syngas must be cooled to ($240°$ to $400°C$). This is done by transferring the heat to a boiler. The cleaned gas expands in a gas turbine and its residual heat is used in the boiler again. Steam from the boiler drives a steam turbine.

Overall efficiency of an IGCC plant increases rapidly with an increase of cleaned syngas temperature up to $350°C$;[31] however, sulphur capturing sorbents decrease in reactivity with temperature, making the development of high temperature ($>300°C$) desulphurization techniques important.

Hot gas clean-up is a major issue of IGCC plants. Currently, ceramics are the material of choice for hot-gas filtration, as they can withstand high temperatures and maintain the extremely high collection efficiency required to prevent erosion of the gas turbine blades. However, ceramic filters are not ideal because they are brittle and susceptible to failure due to mechanical or thermal shock. Current hot-gas filter research is focused on the development of sintered metal filters, which offer resistance to cracking. Preoxidized 2 percent chromium iron aluminide porous metal media is the preferred choice.

Chemical industries have been using IGCC for decades. Out of about one hundred operating gasification plants, there are only five large-scale IGCC plants operating on coal in the utility industry.[32] Five major coal-fired IGCC plants follow:

1. Tampa Electric's Polk Power Station in Florida (Chevron Texaco Gasification Process, 250 MW)
2. PSI Energy's Wabash River Generating Station in Indiana (Global Energy's E-Gas Process, 262 MW)
3. NUON/Demkolec/Willem Alexander IGCC Plant in Buggenum, The Netherlands (Shell Gasification Process, 253 MW)
4. Elcogas/Puertollano IGCC Plant in Puertollano, Spain (Uhde's Prenflo Process, 298 MW)
5. NPRC in Negishi, Japan (342 MW Texaco gasifier, Shell gas clean-up technology)

These plants are of similar design, all using oxygen blown entrained gasifiers. The performance of the five plants is detailed in Table 5.[33,34]

Low-grade coals IGCCs tend to have higher heat rates (or lower efficiencies) than supercritical or ultra-supercritical plants. Overall plant availability is also an issue with IGCC, unless the company takes the expensive route of buying a stand-by gasifier. Furthermore, gasifiers cannot be turned on and off rapidly without impunity due to the heavy refractory in the gasifiers.[32] IGCC, however, has an edge over others when hydrogen, sulfur, and other byproducts of gasification can be sold easily.

Table 5 Operating Commercial Scale IGCC Plants Performance (Adapted from Ref. 33)

	Polk Power (USA)	Wabash River (USA)	NUON/Demkolee (Netherlands)	ELCOGAS (Spain)	NPRC (Japan)
Gas turbine (MWe)	192	192	155	182	N/A
Steam turbine (MWe)	121	104	128	135	N/A
Net power output (MWe)	250	262	253	298	348
Efficiency (%, HHV)	37.5	39.7	41.4	41.5	36
Total operating hours	>25,700 (09/2001)	21,991 (2001)	>23,000 (through 2000)	>6700 (03/2001)	4584 (08/2003)[i]
Coal usage (tons/day)	2,200	2544	2200	2400	N/A
Gasifier availability (%)	84.2[a]	85[b]	50	68[d]	72.3[i]
Power block availability (%)	94.4[a]	89.9[b]	(combined)[c]	84.6[d]	80.7[i]
Emissions					
SO_2 (lb/MWh)	<1.35[e]	1.08[f]	0.44[g]	0.15[g]	<2 ppm
NO_x (lb/MWh)	0.86[h]	1.09[f]	0.7[g]	0.88[g]	<2.6 ppm
Particulates (lb/MWh)	<0.14[e]	<0.10[f]	0.01[g]	0.044[g]	<1.4 mg/m3
Sulphur removal (%)	>98	>97	>99	99.9	99.8

[a]Year 5 operation, ending September 2001.
[b]Year 5 operation in 2000.
[c] Average plant availability in 2000 through September.
[d]2001 operating statistics thought 9/2001.
[e]Reported emissions in 2000.
[f] Average emissions in 2001.
[g] Average emissions reported for 2001.
[h] Average of 14 months of CEMS data.
[i]From gasifier first lit.

Partial Gasification Combined Cycle (PGCC)

Much the same as the IGCC, partial gasification combined cycle (PGCC) plants generate both gas for use in a gas turbine and steam—the difference being in the gasification of the coal. A partial gasifier uses air instead of oxygen to pyrolyze or partially gasify the coal into a lean gas in a pressurized gasifier. The remaining char particles are then captured and fed into a pressurized circulating fluidized bed boiler. The boiler produces steam at supercritical pressure to drive a steam turbine as well as high-pressure hot flue gas containing excess oxygen that helps burn the lean gas in the combustion chamber to drive a gas turbine.

This process is more efficient (>50 percent) than IGCC because the steam turbine inlet temperature is not limited by the exhaust gas temperature of the gas turbine, thus resulting in lower emissions per kW. PGCC also has the potential to have a lower capital cost and higher reliability than IGCC due to the absence of oxygen separation unit, complete gasifier, and the increased simplicity of the system.[35] However, PGCC is not suitable for hydrogen production and is less proven than IGCC as there are no industrial scale plants currently in operation. It can work with a CO_2 sequestration system.

Fuel Cells

Fuel cells are direct energy conversion systems with a very high conversion efficiency. The gasification of coal to produce hydrogen for use in fuel cells offers an efficient means of electricity generation from coal. If a solid oxide fuel cell (SOFC) was run off the syngas produced from coal gasification (Figure 17) and the heat from the fuel cell (SOFCs operate between 850°C and 1000°C)

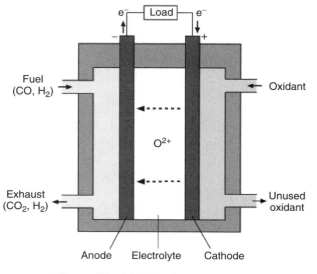

Figure 17   SOFC fuel cell operation.

recovered for use in a Brayton/Rankine combined cycle, then efficiencies of up to 60 percent could be realized.[27]

Fuel cells are a viable option for future power generation, but still need to overcome major reliability obstacles and find ways to decrease the capital and operational costs.

4   POSTCONVERSION CLEAN-UP

Postconversion clean-up is necessary for all types of coal-fired plants to reduce their environmental impact. Control technologies currently exist to remove particulates, NO_2, SO_2, N_2O, and CO_2 from the flue gas. Technologies for the control of heavy-metal emissions are under development. The following sections discuss control options for these pollutants, with the exception of carbon dioxide, which is discussed separately in Section 5.

4.1   Particulates

Particulates in the flue gas include very fine ash and unburnt carbon particles released during the combustion or conversion of coal. The amount of particulates depends on the firing method and ash content of the coal fired. Typically, a pulverized coal (PC)–fired boiler, using very fine (75 percent below 75 microns) coal particles, will see 80 percent of the ash in the coal released as fly ash, with the other 20 percent collected as slag or bed ash at the bottom of the boiler.[36] For a fluidized bed combustor the ratio could be reversed due to the much larger size (75 percent below 6000 microns) of coal particles fired in these types of systems.

There are two common types of control technologies used in coal-fired plants to reduce particulate emissions:

1. Bag filters
2. Electrostatic precipitators (ESPs)

Cyclones are used to collect only large particles and will not be discussed here.

Bag Filter
A bag filter offers a porous barrier to the dust-laden flue gas, capturing the particulate matter while allowing gasses to pass through it (Figure 18). Individual bag filters are arranged in rows in bag houses, where cooled flue gas flows from high to low pressure (keeping the bags inflated) trapping particulates, allowing only the flue gases through. There are two types of dust filtration systems:

1. Depth filtration
2. Surface filtration

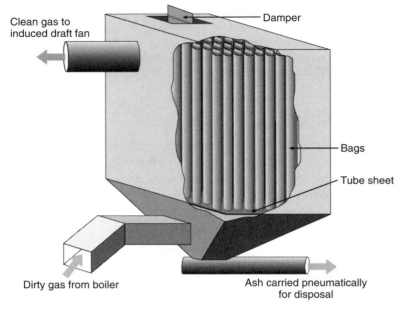

Figure 18   Typical bag house.

In depth-filtration the mesh in the fabric bag is generally not fine enough to capture the smallest particles, which pass through the pores in the mesh. However, as particles pass through, a cake of fly-ash forms on the fibers of the mesh, restricting the pore size, trapping finer particles and at the same time increasing the pressure drop across the bags (Figure 19). Thus, the dust particulates trapped in the filter media facilitate the actual filtering.

In contrast, surface filtration occurs mainly on the surface of the filter media, with little penetration of dust onto the interstices of the filter media because the fabric has extremely small pores. Gor-Tex® and P84 are examples of surface filtration materials.[37] These filters can tolerate higher air-to-cloth ratios and even

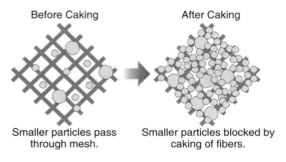

Figure 19   Depth-filtration caking.

a small amount of condensation, which would cause normal bag filters to clog and cease operating. These types of microporous filter materials are still in the testing stage, and wide-scale implementation has yet to take place.

The dust cake formed on either depth or surface filters must be shed through either of the following two methods:

1. Periodical pulse cleaning of the bags when off-line
2. Back-flushing of each bag using high pressure air

Failure to do this will result in clogging and/or rupturing of the bag. The problem with depth-filters is the reduction in filtration efficiency immediately after the cleaning of the bags, when no cake is present. Surface-filtration medium such as Gor-Tex® could allow for collection of finer particles without a drop in the collection efficiency immediately after cleaning.[38]

Bag houses have collection efficiencies of greater than 99 percent even for very small particles. As well, they have the advantage of modular design, reducing costs. The disadvantages of bag houses are that they occupy a great deal of space, as tens of thousands of bag filters are required in large plants to handle the large volume of flue gas. Also, the bags can be harmed by high temperature or corrosive conditions; and as such cannot operate in humid conditions. A large pressure drop resulting in higher auxiliary power consumption is another disadvantage of the bag house. Despite this, bag houses still remain a strong technology of choice for particulate collection.

Air to cloth ratio (cm^3/s per cm^2) is an important design parameter for a bag filter. In a typical coal-fired plant, it may range from 0.75 to 1.1 cm/s with reverse air cleaning and 1.5 to 2.0 cm/s for pulsed jet cleaning bags.[16]

Electrostatic Precipitators (ESP)

Electrostatic precipitators collect particulate matter by applying a static charge to the fly ash particles and drawing them out of the flue gas stream with the opposite charge. As the flue gas passes through a chamber containing anode plates or rods with a potential of 30 to 75 kV,[11] the particles in the flue gas pick up the charge and are collected downstream by positively charged cathode collector plates (Figure 20). Grounded plates or walls also attract the charged particles and are often used for design simplicity. Although the collection efficiency does not decrease, as particles build up on the plates, periodic mechanical wrapping is required to clean the plates to prevent the impediment of the gas flow or the short-circuiting of the electrodes through the built-up ash.

The resistivity of the ash particles is a very important parameter influencing the collection efficiency of ESPs. An ash of high resistivity will not take the charge from the anodes as easily as a less-resistive ash will. If the resistivity is too low, the particles will lose their charge before coming in contact with the collector plates. ESPs can attain efficiencies exceeding 99.5 percent for ash resistivity in the range of 10^4 ohm/cm and 10^{11} ohm/cm. Resistivity of ash is

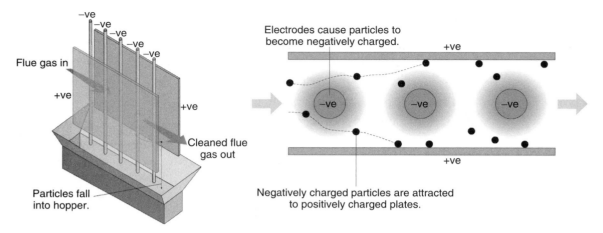

Figure 20 Operation of an electrostatic precipitator.

highest around 200°C and drops sharply above or below this temperature, so the ESP is best operated near this temperature.[16] Higher moisture, sulphur, sodium, and potassium in the coal are favorable for ash collection in an ESP due to their favorable resistivity. Higher calcium and magnesium content in the coal and nonuniform gas distribution in the collector contribute to lower collection efficiencies.

The collection area of the plates also has a direct effect on the collection efficiency, as can be seen in the Deutsch–Anderson equation:

$$\eta = 1 - e^{-\omega A/Q} \tag{11}$$

where

η = efficiency of the collector
A = collection area (m^2)
ω = particle migration velocity (determined from experiments, m/s)
Q = flue gas flow rate (m^3/s)

The main advantages of ESPs are their low operation and maintenance costs, the relatively small footprint (versus bag houses) for large flue gas flows, minimal pressure drop, and constant collection efficiency. A major disadvantage of an ESP is its low-collection efficiency for submicron particles that are of the greatest health concern, as well its sensitivity to the flue gas velocity distribution inside the collector and gas temperature. Even a well-designed ESP can operate poorly if proper conditions are not maintained.

If slagging conditions are present in pulverized coal (PC)–fired plants, molten ash can vaporize and condense in the backpass, forming submicron size inorganic particles, which cannot be captured by conventional particulate control technologies such as ESP or depth-filtration bag houses.

Wet Scrubber

Scrubbers are used to clean the flue gas of gaseous pollutants, as well as particulates. When only particulate matter control is needed, water is used as the scrubbing solution. The water is misted into the flue gas stream, coming into contact with the particulate matter, capturing the particles and condensing out with the particulates.

Wet scrubbers also have high collection efficiencies, upwards of 99 percent, but their major disadvantage is the large increase in possibly toxic wastewater produced.

There are a number of different types of wet scrubbers. All try to maximize the surface area of the scrubber solution and increase the residence time of flue gas in the scrubber to improve efficiency. Three types of scrubber are shown in Figure 21.

4.2 Gaseous Emissions

The oldest method of reducing the harmful effects of gaseous emissions was to disperse them over a wider area using a very tall stack, making the immediate surroundings less toxic while doing nothing for the reduction in acid rain, smog, or ground-level ozone formation. Regulatory authorities in some countries (e.g., India), for that reason, still relate SO_2 capture with stack height.

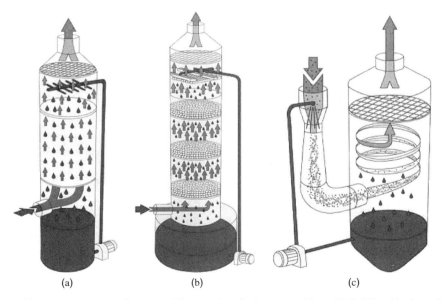

(a) (b) (c)

Figure 21 Various types of wet scrubbers: *(a)* typical wet scrubber, *(b)* fluidized bed scrubber and *(c)* venture scrubber. (Illustration courtesy Forbes—Plastic Tanks and Environmental Technologies.)

Advanced combustion technologies produce less amounts of harmful gases, yet some plants might still need a certain amount of postconversion clean-up of the flue gas.

Sulphur Dioxide

The emission of SO_2 can be reduced either by using fluidized bed combustion in a bed of limestone (see section 3.2) or by scrubbing the flue gas with sorbents for postcombustion removal.

For PC-fired plants, scrubbing is the accepted means for SO_2 reduction. Flue gas desulphurization (FGD) removes the SO_2 from the flue gas stream with a chemical absorbent before it is released into the atmosphere. Sulphur capture is effected using alkaline sorbents such as NH_4, NaOH, or a lime ($CaCO_3$, $CaMg(CO_3)_2$). The ammonia species has the added benefit of capturing NO_x as well. NaOH is generally used as a sulphur carrier and regenerated using lime. Equations (12) to (16) are the reactions involved with respective sorbents:

Sodium hydroxide:
$$2NaOH + SO_2 \rightarrow Na_2SO_3 + H_2O \tag{12}$$

Ammonium:
$$2NaSO_3 + H_2O \rightarrow 2NaHSO \tag{13}$$

Limestone:
$$2NH_4 + SO_2 + O_2 \rightarrow (NH_4)_2SO_4 \tag{14}$$
$$CaCO_3 + SO_2 \rightarrow CaSO_3 + CO_2 \tag{15}$$
$$Ca(OH)_2 + SO_2 \rightarrow CaSO_3 + H_2O \tag{16}$$

Additional limestone is added to the effluent holding tank for precipitation of $CaSO_3$, which can be subsequently oxidized and hydrated with the additional air and water, creating gypsum ($CaSO_4 \cdot 2H_2O$) as seen in equation (17):
$$CaSO_3 + \tfrac{1}{2}O_2 + 2H_2O \rightarrow CaSO_4 \cdot 2H_2O \tag{17}$$

Lime is the most efficient scrubbing agent. Therefore, the subsequent discussion will be based on lime scrubbing, which can be either a wet or dry FGD process.

Wet FGD Process. Wet FGD involves spraying finely ground limestone in aqueous slurry, as a mist into the flue gas stream. The water/calcium sulphite mixture produced is then collected in an effluent holding tank and can be further oxidized and hydrated to form calcium sulphate dehydrate (gypsum), which can be a saleable byproduct if impurity levels are low enough (Figure 22). However, there is a limited market for low-quality gypsum and as such much of the gypsum produced from FGD processes is simply land-filled. However, if ammonia

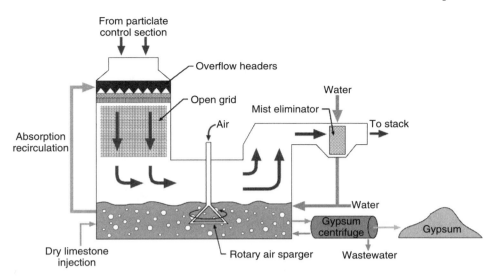

Figure 22   Wet flue gas desulphurization (FGD) process.

is used as the scrubbing agent its byproduct will be $(NH_4)_2SO_4$, which can be used as a fertilizer.

One of the problems with using wet FGD is the formation of sulphuric acid from reactions between the water and SO_2; requiring the use of more expensive stainless steel or other corrosion resistant material. Energy consumption of the wet FGD process ranges from 1 to 4 percent of the total plant-generating capacity.[29] A benefit of using wet FGD is that it also functions as a wet scrubber for particulate matter, as discussed in section 4.1.3.

Dry Scrubbing.   There are two different types of dry scrubbing:

1. Spray-dry scrubbing using sorbent slurries
2. Dry sorbent injection

Spray drying FGD involves atomizing the sorbent slurry in the hot flue gas stream creating a large sorbent-gas surface area for SO_2 absorption (Figure 23). The mixture is then dried in the hot flue gas stream enabling the removal of spent sorbents as a dry powder by means of appropriate particulate control technologies. If a bag filter is used then the flow of flue gases through caked sorbents increases the removal of SO_2 by roughly 10%, due to the increased contact time between the sorbents and the flue gas. Spray drying sulphur removal rates are upwards of 90 percent.[29]

A typical dry injection system injects pulverized lime directly into the hot flue gas stream just after it exits the furnace. As in spray drying, the spent sorbent, along with fly ash, is collected in the particulate control system. Electrostatic precipitators cannot be used here, as they will not collect the sulphate product.

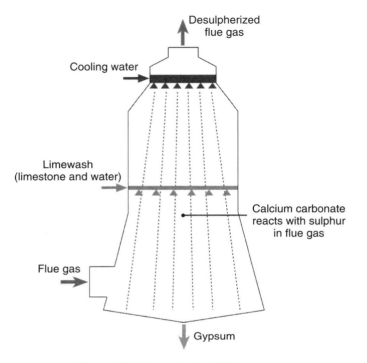

Figure 23   Spray drying FGD process.

The main drawbacks of dry FGD are the high temperatures and high Ca:S ratios required for an adequate level of sulphur capture. Impurities in the fly ash, such as chlorine, could bind to the calcium, lowering the melting temperature of the fly ash and causing fouling problems due to the high temperatures immediately after the furnace.

Benefits of dry FGD over wet FGD include reduced pumping requirements (less water use), elimination of flue gas reheating, reduced corrosion of equipment (little to no acid products are produced), and ease of handling a dry product. However, it requires a relatively high Ca:S ratio, which leads to high alkalinity byproducts that require special handling and disposal. The Ca:S ratio could be reduced if unspent sorbents were recycled.

Nitrogen Oxides

During combustion the nitrogen in both fuel and combustion air could oxidize to form the following three air pollutants:

1. NO (nitric oxide)
2. NO_2 (nitrogen dioxide)
3. N_2O (nitrous oxide)

These compounds cause a number of problems such as:

1. The formation of ground level ozone (NO)
2. The production of photochemical smog (NO_2)
3. The formation of acid rain (NO)
4. The destruction of stratospheric ozone (N_2O)
5. Acting as a greenhouse gas (N_2O)

The oxidation of nitrogen in air (thermal NO_x) below 1,000°C is generally very small. For this reason, fluidized bed boilers operating at around 850°C emit very low amounts of NO_x (Table 6).[42] Furthermore, a number of other reactions occur, especially in the circulating fluidized bed furnace, to reduce the NO_x generated from fuel nitrogen.

The emissions from tangential fired PC boilers (350 to 500 ppm*) are lower than other types of PC boilers, which can have NO_x emissions as high as 1500 ppm.[39] Table 6 shows how the uncontrolled NO_x emissions vary with the types of firing method adopted. The two postcombustion NO_x control technologies are

1. Selective catalytic reduction (SCR)
2. Selective noncatalytic reduction (SNCR)

Both processes use ammonia (NH_4) or urea ((NH_2)$_2$CO) to reduce NO_x back to the stable form of N_2. The reaction is as follows:

$$NO_x + NH_4 \leftrightarrow N_2 + H_2O \qquad (18)$$

Selective Catalytic Reduction (SCR). Selective catalytic reduction (SCR) occurs in a separate reactor (Figure 24) at lower temperatures (300°C to 400°C) in

Table 6 Uncontrolled NO_x Emissions (Adapted from Ref. 39)

Types of Boiler	Fuel Types	Firing Type	Emission Factor in kg NO_x/tonne Coal
Pulverized firing	Anthracite	All types	9
	Bituminous	Front	10.5
		Tangential	7.5
	Lignite	Front	6.5
		Tangential	4
	Residual oil	Vertical firing	12.6 kg/1000 liter
		Tangential firing	5 kg/1000 liter
	Natural gas	All types	8.8 kg/1000 m3
Fluidized bed	Bituminous	Circulating fluidized bed	100–200 ppm

* ppm = parts per million

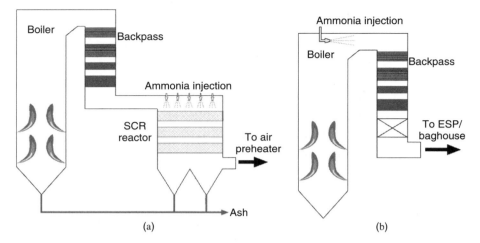

Figure 24 (a) Selective catalytic reduction (SCR) and (b) Selective noncatalytic reduction (SNCR).

presence of a catalyst (i.e., vanadium or a catalyst with zeolites in the wash-coat). To move the reaction forward activated carbon can also be used as a catalyst and it can drive the reaction at even lower temperatures (150°C) with the added benefit of absorbing SO_2.[40] On the downside, SCR has a high installation cost as well as high maintenance and operation costs due to the required periodic replacement of expensive catalysts. Another problem with SCR is that the ammonia injected may not be entirely consumed. The emission of NH_3 through the stack (referred to as ammonia slip) is a common problem for SCR and attention must be paid to the reagent dosage and good mixing must be present in the reaction chamber. A NO_x reduction of 90 percent or more is achievable using a SCR process.[28]

Selective Non-Catalytic Reduction (SNCR). In SNCR, the reducing agent is injected directly into the furnace above the combustion zone in the temperature range of 900°C to 1,100°C without any catalyst. Sufficient residence time in this temperature range and uniform distribution and mixing of the reagent are required for efficient NO_x capture. The capital cost of SNCR is about half of that of SCR due to the absence of catalysts and the separate reaction chamber needed for SCR.[27]

One problem with SNCR is the production of ammonia sulphate (($NH_4)_2SO_4$), which can corrode boiler tubes in the backpass. Tests done on direct ammonia injection (SNCR) in combination with a calcium sorbent have shown reductions of emissions of 85 percent for NO_x and 90 percent for SO_2.[27]

Nitrous Oxide
Nitrous oxide (N_2O) is a major greenhouse gas with a global warming potential 310 times (100-yr basis) higher than that of carbon dioxide, but it cannot be

captured using SCR or SNCR. The intermediate combustion product hydrogen cyanide (HCN) plays an important role in reducing NO into N_2O, which can be reduced back to molecular nitrogen (N_2) if hydrogen radicals are present.[41]

$$HCN + O \rightarrow NCO + H \tag{19}$$

$$NCO + NO \rightarrow CO + N_2O \tag{20}$$

$$N_2O + H \rightarrow N_2 + OH \tag{21}$$

The extent of the reduction of N_2O into N_2 as shown in equation (19) increases with combustion temperature; therefore, a higher-combustion temperature favors lowering the N_2O emissions. As such, the N_2O emissions from PC-fired boilers with combustion temperatures $\sim 1,300°C$ is in the range of 1 to 20 ppm, while that in fluidized bed boilers with combustion temperatures $\sim 850°C$ is in the range of 20 to 200 ppm.[14] It may be noted that NO increases with the combustion temperature, but N_2O decreases with it.

Fossil fuels with a higher-volatile content decrease N_2O formation but increase NO formation. Biomass fuels emit relatively low N_2O.[42]

4.3 Heavy Metals

A total of 11 heavy metals, classified as hazardous air pollutants (HAPs), have been detected in the flue gases of coal-fired power plants. The majority are removed using conventional particulate and gaseous control technologies, with the exception of vaporous mercury.

In 2010 the United States will implement the first emissions controls on heavy metals produced from coal-fired plants. The Clean Air Mercury Rule (CAMR) will implement mercury emissions trading and force a reduction of mercury emissions by 23 percent from 1999 levels and a further reduction to 69 percent of 1999 levels by 2018.[43] These regulations will force coal-fired plants in the United States to either reduce mercury emissions or pay for producing them. This has resulted in a great deal of research into mercury emissions reduction techniques in the United States in recent years.

Mercury in coal appears in three different forms. Solid mercury compound (Hg_p) is effectively removed using existing particulate control technology, with ESPs being most effective due the conductive properties of the compounds. Elemental mercury (Hg^0) and ionic mercury (Hg^{+2}) are in the form of a vapor in the flue gas and represent 90 percent of the mercury emissions. In coals with some chlorine the ionic mercury will be oxidized to produce mercury chloride ($HgCl_2$). Although $HgCl_2$ can be captured by conventional means, it is a very toxic substance and could pose a major waste handing problem. Plants with wet scrubbers have shown a reduction in ionic mercury compounds, but not elemental mercury as it is insoluble in water. SCR may allow a small reduction in the elemental mercury.

The best method to remove the remaining high concentrations of elemental mercury in the flue gas stream is through the injection of a sorbent. One sorbent being investigated is Na_2CO_3, which oxidizes the elemental mercury, making it easy to capture using a wet scrubber.[44] Activated carbon, either injected as powdered activated carbon (PAC) or as fixed-bed granular activated carbon, is an effective absorber of mercury. It has been shown that even without activated carbon injection, 12 percent of the total mercury was absorbed on carbon rich ash.[44] Through the use of sulphur-impregnated active carbon, elemental mercury reductions of more than 99 percent, with a relatively small mass loading, have been observed in laboratory testing.[45]

Testing on existing coal-fired plants has shown that changes to combustion, air preheater, and ESP operation can reduce mercury emissions; however, a trade-off exists between emissions of Hg and NO_x.

4.4   Solid Waste

In Germany and Japan the byproducts from coal combustion must be used in some way, as land-filling of viable products from coal combustion is illegal. In India, where 75 percent of electricity is generated from fossil fuel combustion, 106 million tons of coal combustion residues are produced annually and 73 percent of this put into landfills, showing the need to make use of this waste.[46]

Solid wastes include fly ash, bottom ash, boiler slag, FGD waste, and SCR waste. Ash produced from regular PC-fired boilers has use as an aggregate for concrete, road construction, and general fill. Currently, the U.S. EPA does not classify this waste as a hazardous material, even though it contains trace amounts of heavy metals, which could leach out of a landfill site and cause environmental problems.

The solid residues from a fluidized bed boiler with sulphur capture are composed of spent sorbent $CaSO_4$, which can be hydrated to make gypsum ($CaSO_4 \cdot 2H_2O$), but due to the presence of ash from the coal the quality of the gypsum is reduced and thus the main use for bed ash would be as a low-grade cement or soil remediation. Boilers firing low-sulphur coal without limestone feed produce ash particularly suitable for cement production.

FGD processes using a lime-based sorbent produces gypsum as a byproduct, which can be used in low-quality concrete or wall board manufacture. If the combustion efficiency of the process is low and a large amount of carbon fines remain with the gypsum, then the product is unusable unless the carbon can be removed. In addition, heavy metals and high alkalinity due to unreacted CaO could make solid byproducts toxic and difficult to use or handle. Dry FGD, which uses a higher Ca:S ratio, is more prone to this problem than wet FGD is. If the material has a high alkalinity it could see alternate usage as a soil amendment to neutralize soil acidity.

Integrated gasification combined cycle (IGCC) plants can capture sulphur as either sulphuric acid or elemental sulphur. Sulphuric acid is an important commodity chemical. The major use (60 percent of total worldwide) for sulphuric acid is in the production of phosphoric acid, used for manufacture of phosphate fertilizers, and as tri-sodium phosphate for detergents. It also finds use in sulphate fertilizers, nylon, and lead-acid battery manufacturing. Elemental sulphur has a large number of uses, a few being gunpowder, vulcanized tires, and Epsom salts.

5 CARBON DIOXIDE

Coal produces more CO_2 than any other fossil fuel, and as such, the concern over global warming is a major issue with coal-fired power plants. The easiest means for reducing CO_2 emissions from coal-fired plants is to increase the efficiency of the plant, thus emitting less CO_2 for a given amount of power produced. This alone, may not be enough to curb the rising levels of CO_2 in the atmosphere. Emissions must not only be reduced, but eliminated to curb-rising greenhouse gas levels in the atmosphere. To do this an almost pure stream of CO_2 must be produced and transported to a proper disposal site for permanent storage, as is done for nuclear wastes.

Figure 25 outlines management plans for reducing CO_2 emissions from coal-fired power plants and its sequestration. Reduction in emissions through improved efficiency and co-firing of carbon neutral biomass in coal-fired plants are not discussed here. Carbon dioxide management has three steps:

1. Capture
2. Transportation
3. Disposal (sequestration)

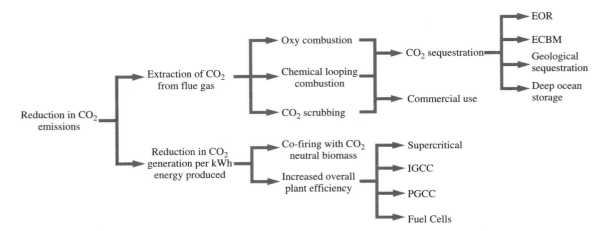

Figure 25 Means for reduction of CO_2 emissions from coal-fired power plants and its sequestration.

5.1 CO$_2$ Capture

The systems for CO$_2$ capture can be grouped under three categories:

1. Precombustion
2. During combustion (oxygen-fired or chemical looping combustion)
3. Postcombustion

Precombustion

Precombustion capture of CO$_2$ involves separation of carbon dioxide from the fuel before it is burnt. Gasification is an example of a precombustion process. Here a synthesis gas is produced in a gasifier through reaction of the coal with air, steam, or oxygen to produce a synthesis gas composed of hydrogen and carbon monoxide.[14] Refer to equations (7) to (10) for gasification reactions.

The CO$_2$ can be separated from the product gas by a suitable separation technique to produce a carbon free hydrogen stream, carrying the energy from the fuel. This hydrogen can be either used in fuel cells or gas turbine engines for energy production.

The flue gas from air-blown coal-fired plants is only 13 to 15 percent carbon dioxide; insufficient for cost-effective transportation and sequestration. The simplest way to obtain an almost pure stream of CO$_2$ is to use a nitrogen-free combustion process, that results in a concentrated stream of CO$_2$ and water (Figure 26). The CO$_2$ could then be easily dehydrated, pressurized, and pumped to the sequestration site.

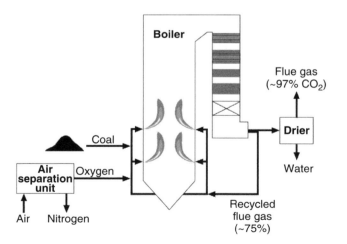

Figure 26 High-concentration carbon dioxide stream produced using oxygen fuel combustion.

Combustion

There are two main methods of separating CO_2 during combustion:

1. Oxy-fuel combustion
2. Chemical looping combustion

Oxy-fuel Combustion. Traditionally, oxygen in air is used as the oxidant for combustion. This produces a flue gas containing large amounts of nitrogen, making it unsuitable for CO_2 sequestration or for commercial use. The nitrogen content in the flue gas can be reduced by adding oxygen to the combustion air,[47] or even eliminated by burning the fuel in pure oxygen. In oxy-fuel combustion plants, oxygen is separated from air and used in the boiler instead of air for combustion. Equations (22) and (23) illustrate the process:

Air combustion:

$$X(0.21O_2 + 0.78N_2) + C_nH_{2m} = mH_2O + nCO_2 + (0.78X)N_2 + \text{Heat} \quad (22)$$

where $0.42X = m + 2n$

Pure oxy-combustion:

$$(m/2 + n)O_2 + C_nH_{2m} = mH_2O + nCO_2 + \text{Heat} \quad (23)$$

In oxy-fuel combustion plants, oxygen is separated from air and used in the boiler instead of air for combustion. Elimination of nitrogen in the oxidant brings with it a number of benefits, including pure carbon dioxide in flue gas:

1. The flue gas is composed of CO_2 and water vapor. After condensing out the water, the flue gas available is pure carbon dioxide that is ready for transport and sequestration or sale.
2. Thermal efficiency of the boiler is enhanced due to reduced flue gas volume and, hence, lower dry flue gas loss.
3. A higher percentage of CO_2 and or H_2O in the flue gas enhances the nonluminous radiation from flue gas and results in a higher specific heat than that of nitrogen, the main dilutent of normal flue gas. This increases the heat transfer rates in the backpass and reduces the required size of heat transfer surfaces.
4. High volumetric and grate heat release rates are obtained, and therefore the size of the boiler furnace could be reduced.[48]

Chemical Looping Combustion. In a chemical looping combustion process, the oxygen for combustion is provided by a metal oxide, which absorbs oxygen from air in a separate reactor. Chemical looping combustion processes result in pure nitrogen in one reactor and pure carbon dioxide, after combustion, in the other.

Figure 27 shows the principle of chemical looping combustion. In reactor A, the fuel, C_nH_{2m} reacts with the oxygen carrier metal oxide, M_yO_x, to produce

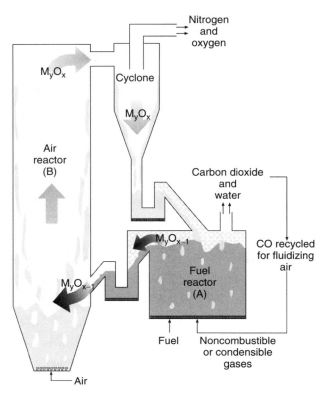

Figure 27   Chemical looping combustion.

CO_2 and H_2O. This exothermic reaction generates the combustion heat. In reactor B, the reduced oxygen carrier M_yO_{x-1} reacts with the oxygen in air, regenerating the oxygen-carrying metal oxide, M_yO_x.[50] The reaction is expressed in equations (24) and (25):

Fuel Reactor (A):

$$(2n + m)M_yO_x[\text{oxygen carrier}] + C_nH_{2m}[\text{fuel}]$$
$$\rightarrow (2n + m)M_yO_{x-1} + mH_2O + nCO_2 + \text{Heat} \qquad (24)$$

Air Reactor (B):

$$(2n + m)M_yO_{x-1} + (n + m/2)O_2 \rightarrow (2n + m)M_yO_x \qquad (25)$$

The major advantage of chemical looping combustion over the oxy-fuel option is that it does not require the expensive oxygen-separation plant. Furthermore, the extra power consumption and hence extra CO_2 generation for oxygen separation is avoided.

Metal oxides such as Fe_2O_3, Mn_3O_4, Cuo, and NiO impregnated on quartz, alumina, or other inert materials act as the oxidant for combustion by transporting the O_2.[51] Johansson et al.[50] studied the reactivity and crushing strengths of

58 different oxygen carriers in a circulating fluidized bed burning methane.[52] Nickel-based particles were most active, followed by copper, manganese, and iron.

Chemical looping combustion is in its early stage of development, and issues around solid fuel combustion and burning in large furnaces are yet to be sorted out.

Postcombustion CO_2 Extraction

Available processes for stripping CO_2 from the flue gas can be classified into three groups:

1. Separation with sorbent/solvents
2. Membrane separation
3. Separation by cryogenic distillation

Separation with Sorbent/Solvents. Currently, the sorbent/solvent absorption method is more advanced and closer to commercialization than other options. With this process, the separation is achieved by passing the flue gas through a reaction chamber in intimate contact with a liquid or solid sorbent that is capable of capturing the CO_2. The sorbent with dissolved CO_2 is then transported to a different vessel where, due to changes in pressure, temperature, or other conditions, it releases the CO_2. The regenerated sorbent is sent back to the first vessel to capture more CO_2 in a cyclic process.

In some processes that use this type of separation, the solid sorbent does not circulate between vessels; the sorption and regeneration are achieved by cyclic changes in pressure. This process is called *pressure swing adsorption* (PSA). In PSA a high-pressure flue gas stream is passed through a porous material, where the CO_2 is preferentially adsorbed. When the pressure is decreased, the CO_2 is de-adsorbed from the porous sorbent for sequestration. Three sorbents recently studied for use in PSA were molecular sieve 13X, natural zeolite ZS500A, and activated carbon.[52]

In amine scrubbing, the flue gas is passed through a large vessel, usually an absorbing tower, and mixed with an amine-based solvent (organic molecule with nitrogen at the core, such as NH_4, NCH_5, etc.), which captures the CO_2 (Figure 28). The CO_2 is then stripped off the amine using large amounts of low-quality heat, producing a stream of concentrated CO_2. This heat is provided by low-temperature steam generated through the burning of extra fuel. The CO_2 must be compressed to 150 bar to facilitate transport and sequestration.[53] The amine is regenerated and fed back to the absorbing tower. One drawback to this would be the possible capture of not only CO_2 but also NO_x and SO_2. Although this is good for boiler emissions, it contaminates the CO_2 stream produced, preventing its commercial use.

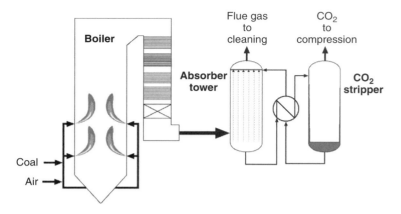

Figure 28 CO_2 capture using an amine scrubber.

Membrane Separation. Gas-separation membranes, used commercially for separation of CO_2 from natural gas, are being considered for flue gas separation. This method relies on differences in physical or chemical interactions between the different substances in a gas mixture and a membrane material that allow one gas to move through the membrane at a faster rate than another. Such a membrane would allow carbon dioxide gas to pass through while excluding the other parts of the flue gas emitted from industry or power plants. If a pressure differential is set up on opposing sides of a membranelike polymeric film, gas transport across the film (permeation) will occur. The CO_2 will permeate through the membrane at a faster rate than the nitrogen or other flue gas components, producing a CO_2-rich stream. Polymers such as cellulose acetate, hollow-fiber polymides,[54] and polypropylene[55] are also being considered as membrane materials.

Separation by Cryogenic Distillation. Distillation is the separation of various gases through the differences in their relative volatility or boiling points. The CO_2-containing flue gas is compressed and cooled below the boiling point of the CO_2. The liquid carbon dioxide is distilled out of the cooled flue gas and separated for storage or use. Carbon dioxide has a relatively high boiling point ($-78°C$ at 1.0 atm) compared to other components of the flue gas, such as nitrogen ($-196°C$ at 1.0 atm). A second distillation of the carbon dioxide stream may be necessary to remove other gasses that also have lower boiling points, such as sulphur dioxide ($-10°C$ at 1.0 atm).

Cost of Separation

The oxy-combustion process needs power for the oxygen separation, requiring additional fuel to be burnt to produce this extra energy. On a fuel-equivalent basis, oxy-combustion requires about 1.32 MWth per tonne of CO_2 avoided, compared to amine scrubbing, which requires 2.8 MWth per tonne of CO_2 avoided.[53] Taking into account an air infiltration ($\sim$3%), 95 percent oxygen purity and extra CO_2 produced for extra auxiliary power the oxy-combustion can avoid about 66 percent of the original CO_2 emission compared to 58% for amine scrubbing, both with 90 percent capture.[53]

5.2 Transportation

For transportation of the captured CO_2 two options are available:

1. *Pipeline transportation*. This is suitable for transportation of large amounts of CO_2 ($>$ 40 million tonnes/yr) over small distances ($<$2500 km).[56]
2. *Use of ships*. This is suitable for small amounts of CO_2 ($<$ few million tonnes/yr) over large distances (overseas).

In the United States, approximately 2,500 km of CO_2 pipeline already supply several million tons of CO_2 per year to EOR projects.[57] Transportation of carbon dioxide through pipelines is done in the gaseous phase to avoid problems associated with two-phase flow as pressure drops in the pipeline. Due to similar chemical properties with propane, transportation of liquid CO_2 in trucks and ships is a readily available transportation option. However, the large amounts of CO_2 produced from coal-fired plants make these methods inadequate, as they can transport only small quantities at a time.

There are a few health concerns with transportation of CO_2 because it is toxic in concentrations above 10 percent and is heavier than air, allowing it to accumulate in low-lying areas. This makes opposition to onshore CO_2 pipelines high; however, offshore pipelines would require longer transport distances, as well as more complex and costly infrastructure.[58]

5.3 Sequestration and Utilization

Carbon dioxide sequestration is the process of keeping anthropogenic CO_2 out of the atmosphere by storing it deep underground. If the CO_2 stream contains other gasses it will make carbon sequestration more costly. These gaseous impurities also increase the minimum miscibility pressure, meaning that the CO_2 stream will have to be pumped to a higher pressure for enhanced oil recovery (EOR) applications, so a nearly pure stream of CO_2 is required. The CO_2 can be stored permanently in either deep onshore/offshore geological formations or at the bottom of the ocean. Several options are available for sequestration of carbon dioxide, as below. Carbon sink management and industrial use are other CO_2 reduction methods.

Underground Sequestration

There are a number of ways to sequester a high-pressure stream of carbon dioxide. For example, it can be pumped deep (~ 800 m) into any of the following:

- Empty oil wells or other geological formations
- Saline aquifers, as done in the Sleipner and Snøhvit projects in Norway
- Active oil wells for enhanced oil recovery (EOR), as done in the Weyburn project in Canada
- Active gas fields, as in the In-Salah project in Algeria (Figure 29).

CO_2 can also be pumped to shallower depths into unrecoverable coal seams where it is adsorbed on coal surface. This also has the potential for enhanced coal bed methane (ECBM) extraction.

Enhanced oil recovery (EOR) is the preferred technology, as it has a value-added effect over geological sequestration and is more proven than ECBM extraction, which does not have any pilot scale testing yet. EOR is being used at the Weyburm oil field in Saskatchewan, Canada, with the pressurized carbon dioxide being pumped 400 km north from the Great Pains synfuel plant in North Dakota at a rate of 5,000 ton/day.

Undersea Sequestration

At atmospheric temperature and pressure, carbon dioxide remains in gaseous state, but it can be turned into a liquid by compression within the temperature range of $-56.5°C$ and $+31.1°C$. Above its critical temperature ($31.1°C$), the gaseous CO_2 can be compressed to a very high density, even exceeding that of water. Thus, CO_2 can be stored under the ocean floor, where the pressure is very high, without of the risk of being released to the atmosphere above. Carbon

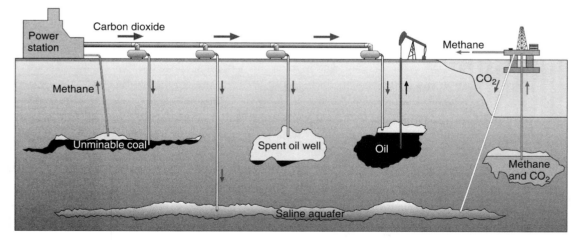

Figure 29   Carbon dioxide sequestration options.

dioxide is soluble in water, but its solubility (0.25 kg CO_2/100 kg water at 20°C and 1 bar pressure) decreases with temperature, pressure, and the salinity of water. At depths below 3,000 m, the pressurized liquid CO_2 is denser than water and will form an undersea lake and not rise to the surface.

There is great concern over this type of CO_2 storage, as there is a chance the CO_2 will dissolve into the surrounding water over the long term, acidifying the water and destroying aquatic life.

Industrial Utilization

Industrial use of CO_2 is the only disposal/sequestration option with direct revenues. With this option, the captured CO_2 is used as a feedstock in chemical processes. Pure CO_2 finds use in the industry as a coolant, to carbonate beverages, and in the pharmaceutical industry as a nontoxic solvent. These uses can absorb only a fraction of the CO_2 generated by even a medium-size coal-fired plant, which generates about 8,000 tons/day. About 100,000 tons of CO_2 are used by U.S. industry annually.[58] Using CO_2 for such purposes does not eliminate the CO_2 emissions as it is still released into the atmosphere in the end.

Carbon Sink Management

The Kyoto Protocol and Marrakesh Accords to the United Nations Framework Convention on Climate change recognize land use, land-use change, and forestry activities—mainly afforestation and reforestation—as a potential means to reduce carbon dioxide concentration in the atmosphere and help developed countries to meet the reduction targets allocated through Kyoto.[59]

5.4 Cost Implications

The additional energy required and the cost involved in CO_2 separation and sequestration are preventing immediate commercial implementation of carbon dioxide separation and its sequestration from coal-fired power plants. The National Energy and Technology Laboratory (NETL) of the United States, along with other groups around the world, are currently studying this issue in the hope of developing viable solutions to reduce the cost of CO_2 separation and sequestration. For example, in a 400 MW net power plant, using oxy-fuel combustion of pulverized coal with cryogenic oxygen separation, the base plant power consumption was 30 MW, while that for CO_2 capture was 102 MW (CO_2 compression = 37 MW, CO_2 transport over 10 miles is greater than 1 MW, storage in an aquifer is 3 MW).[61] Ciferno and Plasynski[61] compared different technologies for CO_2 capture and their costs (Table 7). It is noteworthy that the rise in cost of electricity ranged from 37 percent and 66 percent for different CO_2 capture technologies used. Another study compared the CO_2 capture cost for three technologies and found amine scrubbing to be least expensive (Table 8).[62]

Table 7 Cost of Carbon Capture and Sequestration[a] (Adapted from Ref. 61)

	Amine scrubbing	Aqueous ammonia CO_2 capture	Oxy-fuel PC combustion with cryogenic ASU	Oxy-fuel PC combustion, membrane ASU
Units in Operation	1	4	6	7
Increase in COE %	66	37	64	40
CO_2 avoidance cost, $/tonne	42	24	42	25
Increase in capital cost, %	64.6	43.7	77.8	46.3
Efficiency[b], % (HHV)	29	34	30	32
Auxiliary load, MW	85	78	172	125

[a]for a double reheat supercritical plant with 400 MW net power, 80% capacity factor firing bituminous #6 coal, 90% CO_2 capture, compressed to 2200 psia, transported 10 miles and stored in a saline formation
[b]Base case efficiency and auxiliary load are 41% and 21 MW, respectively

Table 8 Techno-economic Comparison of CO_2 Capture Technologies (From Ref. 62)

Fuel		Bituminous	Subbituminous	Lignite	Lignite	Lignite
Technology		Gasification	Gasification	Gasification	Amine Scrubbing	Oxyfuel Combustion
COE*	$/MWh	107	97	131	116	152
CO_2 emitted	Tonne.MWh	0.116	0.111	0.182	0.06	0.145
CO_2 avoided	Tonne/MWh	0.65	0.74	0.71	0.82	0.74
Cost of CO_2 avoided	$/Tonne	47	52	88	57	112
Capacity	MW net	594	437	361	311	373
Net heat rate	kJ/kWh	11410	13,810	13,240	12,530	14,880
Unit cost	$/kW net	3000	3400	4400	4400	6200

*Cost of electricity for 90% capacity factor

Richards and Stokes[63] analyzed data from several countries ranging from India to the United States for the cost of carbon sequestration through forestation, it being highly country and geography specific. In general, the cost is lower than that for sequestration under the sea or ground.

The sequestration cost depends to a great extent on the pipeline length, sequestration site, geological formation, amount of carbon dioxide sequestered, and so on. Cudnik[64] predicted a cost of $1 to $70/ton CO_2 for 100 to 3,500 Mt of CO_2, but it went sharply up to about $700 when the requirement exceeded 3,500 Mt, as capacities were not available at the site. CCPC[62] studied three specific sites in Canada and predicted a cost of $38/ton CO_2 for enhanced oil recovery sequestration 200 km away, $10/ton CO_2 for sequestration for enhanced coal bed methane extraction 100 km away and \$4/ton CO_2 for geological sequestration 75 km away. Another study for a 500 MW plant in Ontario predicts a cost \$7.5 to \$11.5/ton CO_2 for sequestration at 1,000 m depth in sea 150 km away.[65] The sequestration cost was 10 to 30 percent of the total cost of CO_2 capture and sequestration.

6 CONCLUSION

Coal currently supplies 38 percent of the world's electricity needs, and this number is expected to grow by 1.4 percent per year over the next 30 years.[1] This growth in coal usage could occur in an environmentally responsible way, because available *cleaner coal technologies* can meet even the most stringent emission requirements with only a modest increase in the price of electricity.

Utilities have made great improvements in the reduction of particulates, SO_2, and NO_x emissions from coal-fired power plants, and this will likely continue to improve in the coming decades. The improvements will be driven by the economic disincentives associated with polluting, brought about by government legislation. In the near term, supercritical plants with PC or CFB firing will dominate coal-based power generation, followed by IGCC. More advanced technologies like partial gasification combined cycle and fuel cell may take some time to come to the mainstream utility market.

Reducing and eventually eliminating carbon dioxide emissions will be a very challenging, although not impossible, task for the coal industry. The positive news is that the utility industry has been able to meet every emission challenge in the past, and this new challenge should be no different. The need to avoid potentially drastic, irreversible changes to our climate is imperative and is reflected in the following quote: *"There is broad agreement within the scientific community that amplification of the Earth's natural greenhouse effect by the build-up of various gases introduced by human activity has the potential to produce dramatic changes in climate."**

In 1990, the CO_2 concentration in the atmosphere was 350 ppm.[66] By 2006, the concentrations rose to about 380 ppm, a dramatic 8.5 percent increase in 16 years. Technologies for separation and disposal of the CO_2 from coal-fired plants are available to arrest this dramatic rise in CO_2 concentration within a reasonable cost. *"Only by taking action now can we ensure that future generations will not be put at risk."**

REFERENCES

1. IEA, World Energy Outlook (2006).
2. U.S. Department of Energy, FutureGen—Tomorrows Pollution Free Power Plant, www.fossil.energy.gov/programs/powersystems/futuregen/. Accessed 2003.

---

* Statement by 49 Nobel Prize winners and 700 members of the National Academy of Sciences, 1990.

3. European Commission, Outline Concept and Tentative Structure for the Technology Platform—Zero Emission Fossil Fuel Power Plants, Advisory Council Meeting (September 9, 2005), Report noTP-ZE-AC-4/2005, http://ec.europa.eu/research/energy/nn/nn_rt/nn_rt_co/article_2268_en.htm.

4. M. Eastman, *"Clean coal power initiative"*, 2nd US–PRC conference on Clean Energy, Washington, NETL, U.S. Department of Energy, November 2003.

5. Ontario Ministry of the Environment, "Air Quality in Ontario—2004 Report," Environmental Monitoring and Reporting Branch, 2004.

6. FDA, *What You Need to Know about Mercury in Fish and Shellfish*, FDA, Washington D.C., March 2004.

7. J. B. Calvert, "Mercury: The Lore of Mercury, Especially Its Uses in Science and Engineering," University of Denver Web site, http://www.du.edu/~jcalvert/phys/mercury.htm# Pois. Accessed May 2004.

8. National Mining Association, "Clean Coal Technology: Current Progress, Future Promise," www.nma.org/technology/environmental.asp (2002–2007).

9. World Coal Institute Web site, http://www.worldcoal.org/pages/content/index.asp? PageID=19. Accessed 2006.

10. IEA, "Clean Coal Technologies," home page, IEA Clean Coal Center website, http://www.iea-coal.org.uk/content/default.asp?Pageld=62&Languageld=0, 2007.

11. N. Berkowitz, *An Introduction to Coal Technology* (2nd ed.), Academic Press Inc., 1994.

12. H. Termuehlen and W. Emsperger, *Clean and Efficient Coal-fired Power Plants*, ASME Press, New York, 125, 2003.

13. P. Basu, "An Arrangement for Conversion of Existing Fossil Fuel Fired Boiler into Circulating Fluidized Bed Firing," Canadian patent no 2159949, November 5, 2005.

14. P. Basu, *Combustion and Gasification in Fluidized Beds,* Taylor & Francis, CRC Press, Boca Raton, Florida, 2006.

15. NETL, "Clean Coal Technology Roadmap," Department of Energy, the Electric Power Research Institute and the Coal Utilization Research Council, http://www.netl.doe.gov/technologies/coalpower/cctc/pubs/CCT-Roadmap.pdf, 2006.

16. S. C. Stultz, (ed.), *Steam*, 41st ed., Babcock & Wilcox, Barberton, pp. 33-7-10, 2005.

17. NRCan, CANMET Energy Technology Center website, Natural Resources Canada, http://www.nrcan.gc.ca/es/etb/cetc/cetc01/htmldocs/home_e.htm, 2007.

18. S. J. Goidich, S. Wu and Z. Tan, "Design Aspects of the Ultra-Supercritical CFB Boiler," Proceedings of the International Pittsburgh Coal Conference, Pittsburgh, PA, September 12–15, 2005.

19. J. M. Beér, "Combustion Technology Developments in Power Generation in Response to Environmental Challenges," *Progress in Energy and Combustion Science*, **26**, 301–327 (2000).

20. S. Azuhata, "Advanced Clean Coal Technology." *IEEE* (2000).

21. P. Basu and P. K. Halder, "A New Concept for Operation of Pulverized Coal Fired Boiler Using Circulating Fluidized Bed Firing," *Transactions of the ASME*, pp. 626–630, October 1989.

22. P. Basu and J. Talukdar, *Revamping of a 120 Mwe Pulverized Coal Fired Boiler with Circulating Fluidized Bed Firing*, Proceedings of the 16th International Fluidized Bed

Combustion Conference, ASME, Donald W. Geiling (Editor), Reno, Nevada, May 13–16, 2001.

23. S. Kavidass, D. J. Walker, and G. S. Norton Jr., *"IR-CFB Repowering: A Cost-Effective Option for Older PC-Fired Boilers,"* Power-GE International 1999, November 30–December 2, 1999.

24. R. Viswanathan, J. F. Henry, J. Tanzosh, G. Stanko, J. Shingledecker, B. Vitalis, and R. Purgert, "U.S. Program on Materials Technology for Ultra-supercritical Coal Power Plants," *Journal of Materials Engineering and Performance*, **14**(3), 281–292 (June 2005).

25. G. Sormani, and G. Moscatelli, *"AVEDORE no. 2—The Most Advanced Steam Turbine Generator in the World,"* International Exhibition & Conference for the Power Generation Industries—Power-Gen, 81, 1997.

26. G. Sormani, and G. Moscatelli, *"AVEDORE no. 2—The Most Advanced Steam Turbine Generator in the World,"* International Exhibition & Conference for the Power Generation Industries—Power-Gen, 81, 1997.

27. M. K. MacRae, *"New Coal Technology and Electric Power Development,"* Canadian Energy Research Institute, 1991.

28. B. Davidson, "Clean Coal Technologies for Electricity Generation," *Power Engineering Journal*, **7**(6), pp. 257–263 (December 1993).

29. IEA, Review, *"China Makes a Start on Advanced 'Clean Coal' Generation,"* News, p. 12 (February 2005).

30. D. R. Simbeck, N. Korens, F. E. Biasca, S. Vejtasa, and R. L. Dickenson, *"Coal Gasification Guidebook: Status, Applications, and Technologies,"* TR-102034, Final Report, Prepared for Electric Power Research Institute, December 1993.

31. K. Jothimurugesan, S. K. Gangwal, R. Gupta, and B. S. Turk, *"Advanced Hot-Gas Desulphurization Sorbent,"* Advanced Coal-Based Power and Environmental Systems '97 Conference, July 1997.

32. S. Blankinship, *"Amid All the IGCC talk, PC Remains the Go-to-guy,"* "Power Engineering", 16–22 (April 2006).

33. J. Ratafia-Brown, L. Manfredo, J. Hoffmann, and M. Ramezan, *"Major Environmental Aspects of Gasification-Based Power Generation Technologies,"* Final Report, Gasification Technologies Program National Energy Technology Laboratory, U.S. DOE, December 2002.

34. World Coal Institute, "Clean Coal: Building a Future through Technology," World Coal Institute Online publications, http://www.worldcoal.org/pages/content/index.asp? PageID =36, July 2004.

35. A. Robertson, Z. Fan, D. Horazak, R. Newby, H. Goldstein, and A. C. Bose, "2nd Generation PFB Plant with W501G Gas Turbine", Proceedings of FBC2005, 18th International Conference on Fluidized Bed Combustion, 2005.

36. P. Basu, C. Kefa, and L. Jestin, *"Boilers & Burners—Design and Theory,"* *Springer*, "New York", p. 57, 2000.

37. G. E. Tooker, "Troubleshooting Dust Collection Systems," Air Control Science Inc., www.aircontrolscience.com/uploads/publications/, (1996).

38. S. J. Miller, G. L. Schelkoph, G. E. Dunham, K. Walker, and H. Krigmont, *Advanced Hybrid Particulate Collector, A New Concept for Air Toxics and Fine-Particle Control,*

Topical Report prepared for Program Research and Development Program, U.S. DOE, 2002.

39. Environmental Protection Agency of USA, *"Control Techniques for Nitrogen Oxide Emissions from Stationary Sources,"* Revised 2nd ed., EPA-450/3-83-002. Research Triangle Park, NC, 1983.

40. K. Knoblauch, E. Richter, and H. Juentgen, "Application of Active Coke in Processes of SO_2 and NO_x Removal from Flue Gasses," *Fuel*, **60**(9), pp. 832–838 (September 1981).

41. L. E. Amand and S. Andersson, "Emissions of Nitrous Oxide from Fluidized Bed Boilers," *Proceedings of the 10th International Conference on Fluidized Bed Combustion*, 1989.

42. B. Leckner, M. Karlsson, M. Mjornell, and U. Hagman, "Emissions from a 165 MWth Circulating Fluidized Bed Boiler," *J. of Inst. of Energy*, **65**(464), pp. 122–130 (1992).

43. J. Winters, "News and Notes—Quashing Quicksilver," *Mechanical Engineering*, **127**(11), (November 2005).

44. Y. Tan, R. Mortazavi, B. Dureau, and M. A. Douglas, "An Investigation of Mercury Distribution and Speciation During Coal Combustion," *Fuel*, **83**, 2229–2236 (2004).

45. R. Yan, D. T. Liang, L. Tsen, Y. P. Wong, and Y. K. Lee, "Bench-Scale Experimental Evaluation of Carbon Performance on Mercury Vapour Adsorption," *Fuel*, **83**, 2401–2409 (2004).

46. P. Asokan, M. Saxena, and S. R. Asolekar, "Coal Combustion Residues—Environmental Implications and Recycling Potentials," *Resources, Conservation and Recycling*, **43**, pp 239–262 (2005).

47. R. Hugh, L. Jia, Y. Tan, E. J. Anthony, and A. Macchi, *"Oxy-fuel Combustion of Coal in a Circulating Fluidized Bed Combustor,"* Proc. 19th International Conference on Fluidized Bed Combustion, Vienna, Ed. Winter, F., Part—I, May 22–24, 2006.

48. J. Saastamoinen, A. Tourunen, T. Pikkarainen, H. Hasa, J. Miettinen, T. Hyppanen, and K. Myohanen, *"Fluidized Bed Combustion in High Concentrations of O_2 and CO_2,"* *Proceedings 19th International Conference on Fluidized Bed Combustion*, Vienna, Ed. Winter, F., Part—I, May 22–24, 2006.

49. IEA. "Clean Coal Technologies," IEA Briefing Paper #83, November 2004.

50. M. Johansson, T. Mattison, and A. Lyngfelt, *Comparison of Oxygen Carriers for Chemical-looping of Methane-rich Fuels*, Proceedings 19th International Conference on Fluidized Bed Combustion, Vienna, Ed. Winter, F. Part—I, May 22–24, 2006.

51. J. Adanez, P. Gayan, J. Celaya, L. F. de Diego, F. Garcia-Labiano, and A. Abad, "Behavior of a Cuo-Al2O3 oxygen carrier in a 10kW Chemical-looping combustion plant," Proceedings of the 19th International Conference on Fluidized Bed Combustion, Vienna, May 22–24, 2006.

52. R. Siriwardane, M. Shen, E. Fisher, J. Poston, and A. Shamsi, *"Adsorption and Desorption of CO_2 on Solid Sorbents,"* Prepared for U.S. Department of Energy, National Energy and Technology Laboratory, 2002.

53. D. J. Singh, E. Croiset, P. L. Douglas, and M. A. Douglas, "CO_2 Capture Options for an Existing Coal Fired Power Plant: O_2/CO_2 Recycle Combustion vs. Amine Scrubbing," First National Conference on Carbon Sequestration, National Energy and Technology Laboratory, May 14–17, 2001.

54. K. Nagao, N. Booker, A. Mau, J. Hodgkin, S. Kentish, G. Stevens, and P. Geertsema, *"Gas Permeation Properties of Polyimide/epoxy Composite Materials,"* 22nd ACS Fall National Meeting, Chicago, Aug 26–30, 2001.

55. Z. Zhikang, W. Jianli, C. Wei, and X. Youyi, "Separation and fixation of carbon dioxide using polymeric membrane contactor." First National Conference on Carbon Sequestration, U.S. Department of Energy, NETL, May 14–17, 2001.

56. B. Metz, O. Davidson, H. Coninck, M. Loos, and L. Meyer, "Carbon Dioxide Capture and Storage." Intergovernmental Panel on Climate Change, Special report. ISBN 92-9169-119-4, September, 2005.

57. R. W. Luhning, J. H. Glanzer, P. Noble, and H.-S. Wang, "Pipeline Backbone for Carbon Dioxide for Enhanced Oil Recovery in Western Canada," *Journal of Canadian Petroleum Technology*, **44**(8), 55–58 (August 2005).

58. R. Svensson, M. Odenberger, F. Johnsson, and L. Stromberg, "Transportation Systems for CO_2—Application to Carbon Capture and Storage," *Energy Conversion and Management*, **45**, 2343–2353 (2004).

59. Ahn, SoEun, *"An econometric analysis on the costs of carbon sequestration in Korea,"* Korea Environment Institute (2005).

60. Z. Xu, J. Wang, W. Chen, and Y. Xu, "Separation and Fixation of Carbon Dioxide Using Polymeric Membrane Contactor," First National Conference on Carbon Sequestration, U.S. Department of Energy, NETL, May 14–17, 2001.

61. J. P. Ciferno and S. I. Plasynski, "System Analyses of CO_2 Capture Technologies Installed on Pulverized Coal Plants," Technical Facts, U.S. DOE, National Energy Technology Laboratory, Sequestration, December 2005.

62. Canadian Clean Power Coalition, "Summary Report on the Phase I Feasibility Studies," May (2004).

63. K. R. Richards and C. Stokes, "A Review of Forest Carbon Sequestration Cost Studies: A Dozen of Years of Research," *Climate Change*, **63**, pp. 1–48 (2004).

64. R. Cudnik, "Managing Climate Change and Securing a Future for the Midwest's Industrial Base," Battelle Columbus Operations, Regional Carbon sequestration Partnership Meeting, Nov 3–4, 2003.

65. A. Shafeen, *"CO_2 sequestration opportunities for Ontario,"* Fossil Fuels & Climate Change Group, CEPG, CANMET Energy Technology Centre, Ottawa, (2003).

66. NOAA, "Atmospheric CO_2 at Mauna Loa Observatory," ESRL Global Monitoring Division, web site: http://www.cmdl.noaa.gov/ccgg/trends/co2_data_mlo.php, 2006.

CHAPTER 9

USING WASTE HEAT FROM POWER PLANTS

Herbert A. Ingley III
University of Florida
Gainesville, Florida

This chapter discusses several examples of integrating power production with the utilization of the associated waste heat to accomplish some other function, such as space heating, domestic water heating, cooling, steam production, or process heating. In addition to several domestic applications of combined heat and power, two new innovative systems for water purification and improved thermoelectric power production are reviewed.

It has become common practice to combine heat recovery with power production, but generally this has been accomplished on fairly large scale (60 megawatts and up). During the 1970s, when the United States had its eyes opened to the fact that our energy resources are finite and global influences can and will affect these energy resources, concepts such as cogeneration and combined cycle power production gained widespread attention.

Conventional Rankine steam power plants (Figure 1) burn fossil fuels to produce high-pressure steam that is expanded through turbines to drive electric generators to produce electricity. This electric power is then transmitted over transmission lines and via transformers, and these lines are distributed to the utility's customers. When the efficiencies of combustion, generation, and transmission are combined, the overall efficiency ranges from 30 to 35 percent. In the process, waste heat and carbon dioxide are rejected to the environment. These large Rankine power plants require a significant period of time to start up and are not as flexible when it comes to responding to changes in peak power requirements. With the addition of gas turbine machines, power plants could then operate these machines as peaking units and respond to changes in peak power demands. However, these early turbine units were not as efficient as the base steam power unit. These turbine units have since seen significant improvements, and they operate at much higher efficiencies. Combining the operation of these turbine units with Rankine cycles exploits the waste heat from the turbine to produce additional power from the steam cycle. The overall efficiencies of these combined cycles with transmission and distribution losses now approach 50 percent. However, continued increases in energy costs and the uncertainties of future energy

Figure 1 A coal-fired power plant and associated distribution system.

resources are promoting newer technologies to gain further improvements in waste energy recovery associated with power production. By moving the source of power production closer to the end-use facility and integrating the use of waste heat with specific energy requirements at the facility, additional energy savings can be accomplished. The savings come from reduced electrical distribution losses and from user-specific applications of the waste heat. Additional benefits come from the overall reduction of the production of the global warming carbon dioxide.

In order to distinguish these new technologies, it seems we need to assign a new name, and in fact an acronym, to this technology—CHP, or combined heat and power, a form of co-generation. This onsite production of electricity is gaining popularity, especially with industries that have uses of waste heat. With CHP systems, there is considerable flexibility in which type of power system is used. Examples include turbines, combined cycle technology, microturbines, fuel cells, and internal combustion engines. Successful implementation of CHP systems depends on several conditions. In addition to being able to utilize waste heat, the CHP candidate must also be able to market excess electrical power. Many states have restructured their electric industries to encourage the purchase of power from CHP systems.

The heat recovered from the production of power in a CHP system can be used for a variety of purposes, including process heating, space heating and cooling, and dehumidification or regeneration of dehumidification systems. The power produced by CHP systems is used near the point of production, resulting in much lower transmission and distribution losses compared to conventional power systems. The overall efficiencies resulting from CHP can be more than twice

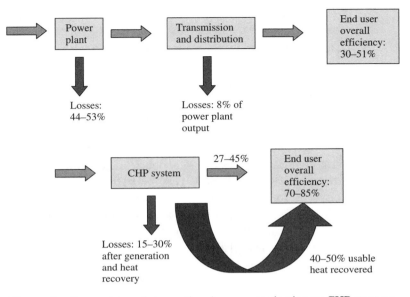

Figure 2   Comparison of conventional power production to CHP systems.

the conventional power system (Figure 2). This also implies less environmental impact in terms of the production of global warming gases and thermal pollution of water resources. This distribution of power can also result in a more secure power supply, with the conventional utility grid backing up the CHP system.

The American Council for an Energy Efficient Economy (ACEEE) reports on its Web site that there are approximately 56,000 megawatts of CHP electric generation operation in the United States. Compare this to the less than 10,000 megawatts of CHP electric generation reported in 1980. The primary applications have been in the chemical, petroleum refining, and paper industries. As the technology of CHP has improved, smaller CHP systems are finding application in other industries such as the food, pharmaceutical, light manufacturing, commercial, and institutional buildings.

ACEEE reports that in 1999, CHP accounted for 7 percent of the United States electricity generation capacity and 310 billion kilowatt-hours of electrical generation. Estimates are that the capacity of CHP power production will continue to grow, with a DOE goal of 92 gigawatts by 2010. With CHP systems reporting efficiencies in the range of 68 to 90 percent, the CHP technology provides an energy-efficient alternative to conventional power (see Figure 2). In addition, ACEEE reports that nitrogen oxide emissions from CHP facilities are one-tenth of those from conventional systems with the same capacity. The State and Local Climate Change Program and EPA predict that high-efficiency CHP could reduce greenhouse-gas emissions by more than 70 million metric tons of carbon equivalent (MMTCE) by the year 2010.

In spite of the optimistic vision for CHP indicated by its current successes, there are several barriers to the further application of this technology. These barriers include a lack of national standards for the interconnection of CHP power systems with conventional power grids, resulting in some utilities requiring expensive studies and equipment before these interconnections are made; penalties in the form of expensive back-up power rates; inconsistencies in depreciation schedules and actual equipment service life; and the lack of awareness of CHP system potential by facility managers.

Other applications of CHP under study include the use of commercially available fuel cells (a PC25 phosphoric acid indirect fuel cell designed and manufactured by ONSI, now UTC) for decentralized power generation with heat recovery for space heating and cooling and domestic hot water.[1] Five 200 kW PC 25 C fuel cells provide a grid-independent/grid-parallel generation system for the Anchorage Processing and Distribution Center in Anchorage, Alaska. This system provides all the electrical power for the facility and most of its thermal needs. Energy cost savings are estimated to be $350,000 per year. Lawrence Berkeley National Laboratory reports that besides the 40 percent electrical efficiency of the natural gas–driven PC 25 fuel cells, 44 percent of the fuel is converted into heat that can be used for space heating, domestic hot water, and cooling through the use of an absorption chiller. Researchers there estimate that the 200 kW fuel cell, combined with a single-stage absorption chiller with a COP of 0.7, can produce 47 tons of refrigeration. The newer version of the fuel cell (PC 25 C) produces waste heat in the temperature range of 250°F, which can be used to operate a two-stage absorption chiller with a COP of 1.2. Another significant application of CHP is the use of a CHP system to produce ethanol where there is a near-perfect match between the CHP's electric and steam production and that of the ethanol plant.[2]

Another application of power plant waste heat that is receiving international attention is a new desalination technology that uses the waste heat from a power plant to purify water. Many coastal communities in the United States and many countries abroad are experiencing a shortage of potable water. These communities are seeking and implementing technologies, such as reverse osmosis plants to desalinate water, to meet their potable water demands. Lior indicates that water desalination has increased in use over the last four decades to the point that over 5.3×10^9 gallons (20 million m³) of fresh water are being produced by over 10,000 land-based plants.[3] Y. Li et al. reports for 13 countries that 2002 water prices range from approximately $1.75/1,000 gallons to approximately $6.75/1,000 gallons.[4] Researchers at the University of Florida are studying an innovative method of desalination that uses the waste heat from power plants as a source of heat.[4] The diffusion driven desalination (DDD) method proposed by the researchers is illustrated in Figure 3. Warm water is evaporated into a low-humidity air stream, and the vapor is then condensed to produce distilled water. Even though this process has a low distilled water output to feedwater

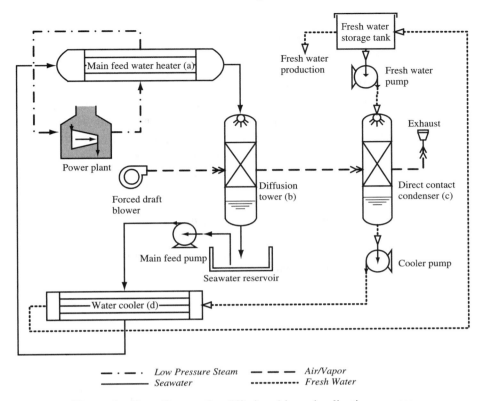

Figure 3 Flow diagram for diffusion driven desalination process.

conversion efficiency, the process has been shown to be cost-effective when coupled with the low-grade waste heat associated with power plants. The University of Florida researchers report that at optimum operating conditions with a high temperature of 122°F (50°C) and sink temperature of 77°F (25°C), an air mass flux of 0.307 lbm/ft²-s (1.5 kg/m²-s), air to feedwater mass flow ratio of 1 in the diffusion tower, and a fresh water to air mass flow ratio of 2 in the condenser are realized. Operating at these conditions yields a fresh-water production efficiency of 0.035 (mass of fresh water to mass of salt water input) and electric energy consumption rate of 0.001kW-h/lb fresh water (0.0022 kW-h/kg fresh water). The researchers evaluated several scenarios. A brief economic analysis is presented to demonstrate the added value provided by using a DDD facility to produce fresh water using waste heat from a thermoelectric power plant.

Y. Li et al. provide an example of a 100 MW thermoelectric power plant with a thermal efficiency of 40 percent, retrofitted with a DDD plant.[4] The total input energy to the power plant would be 250 MW and the waste heat generated 150 MW. With a condenser operating at 9.7 cm Hg pressure, there is approximately 150 kW of energy available at 122°F (50°C) from the low-pressure condensing steam. A DDD plant would have the potential of producing as much as 1.14 million gallons of fresh water per day. The footprint for the DDD plant is estimated to be approximately 0.5 acres.

Another group of researchers at the University of Florida under the leadership of William Lear is researching an innovative power plant configuration that uses the waste heat associated with a thermoelectric process. The High Pressure Regenerative Turbine Engine (HPRTE) is a novel power plant that has the potential for high efficiency (also at part power), in a highly compact, low-cost form. The system inherently produces extremely low emissions, even on liquid fuels, while simultaneously producing power, cooling, heating, and fresh water. The very low emissions occur with little fuel sensitivity, so that hydrogen, methanol, low BTU syngas, and others could be used effectively. The efficiency tends to remain nearly constant at part power, making this system particularly attractive for applications requiring load following, such as distributed generation and transportation.

Figure 4 shows a diagram of the combined-cycle thermodynamic processes in the HPRTE. The overall system consists of a gas turbine subsystem, coupled

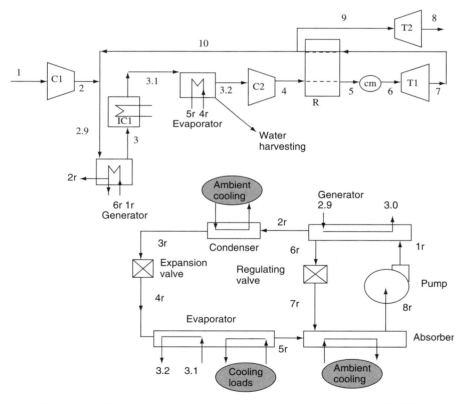

Figure 4 The combined-cycle thermodynamic processes in the HPRTE.

via heat exchange to a vapor absorption refrigeration unit. The novelty lies in the gas turbine, which is a semiclosed Brayton cycle, and the coupling of the two subsystems in the intermediate stages of the gas turbine, rather than at the entrance and exit. The semiclosed gas turbine may be considered to be a turbocharged, intercooled, recuperated gas turbine, in which the inlet air flow is just sufficient to support combustion. This not only enhances compactness and improves part-power efficiency, but also leads to the burner operating in the mild combustion regime, significantly lowering NO_x CO, and unburned hydrocarbon emissions.

The author of this chapter has also been involved in researching processes for utilizing waste heat from power production processes to enhance the overall efficiency of fuel utilization. For the last few years, the Alternative Energy Research group has been researching an ammonia–water combined cycle at the University of Florida's Solar Energy and Energy Conversion Laboratory. This cycle can be used as a bottoming cycle for conventional power plants to produce additional electrical power and refrigeration. The operating temperature for this cycle also lends itself to using solar energy as the primary source of energy.

Figure 5 illustrates the flow diagram for the ammonia–water combined cycle. The fluid leaves the absorber at state 1 as a saturated solution at the cycle low pressure with a relatively high ammonia concentration. It is pumped to the system high pressure (state 2) before traveling through the recovery heat exchanger, where it absorbs heat from the weak solution returning to the

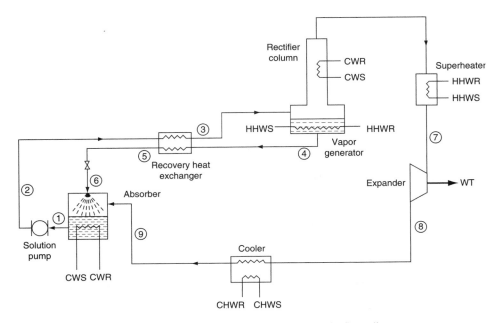

Figure 5 The ammonia/water combined cycle flow diagram.

absorber. The solution is then partially boiled in the vapor generator by the heat source, producing saturated ammonia vapor and relatively weak concentration ammonia–water saturated liquid. The weak solution leaves the vapor generator at state 4 and rejects heat to the high concentration stream before it is throttled to the system low pressure and sprayed into the absorber. The rectifier cools the saturated ammonia vapor to condense out any remaining water. The vapor is then superheated to state 7 and expanded to produce work. The subambient exhaust vapor (state 8) provides refrigeration before returning to the absorber, where it is reabsorbed into the weak solution. The heat of condensation is rejected to the low-temperature source and the cycle repeats.

A small breadboard prototype of this system has been built and studied in some depth.[5] The researchers are now in the process of constructing and evaluating a 5 kW prototype. The prototype has been designed to utilize heating hot water at 180°F (82°C) as a heat source to the vapor generator. The cooling requirements of the absorber and rectifier for the process are met using a conventional cooling tower supplying cooling water at 85°F (29.4°C).

As illustrated by the variety of processes covered in this chapter, there are many new and innovative systems currently in use or in the research and proto-type stage that are seeking to maximize the total efficiency of our fuel consumption to produce power. Many of these new systems are applicable to distributed power scenarios and can be applied to a wide range of facilities, including commercial, institutional, and industrial. In addition to providing much higher levels of efficiency in terms of energy use, these systems will also lead to reduced emissions of global warming gases. The future of systems carrying the names of cogeneration, combined heat and power, total energy systems, or even trigeneration will continue to make a strong contribution to our energy conservation efforts.

REFERENCES

1. "Waste Heat Integration of Electrical Power Generation provided by Fuel Cells," Lawrence Berkeley National Laboratory, Environmental Energy Technologies, Indoor Environment Department, Energy Performance of Buildings Group report, March 11, 2004.
2. Combined Heat and Power Partnership with EPA Web site.
3. Noam Lior, "Water Desalination," *The CRC Handbook of Mechanical Engineering*, 2nd ed., CRC Press, Washington, D.C., 2005, section 20.6.
4. Y. Li, J. F. Klausner, and R. Mei, "Performance Characteristics of the Diffusion Driven Desalination Process," *Science Digest, Desalination* **196** 188–209 (2006).
5. G. Tamm, D. Y. Goswami, S. Lu, A. Hasan, "A Novel Combined Power and Cooling Thermodynamic Cycle for Low Temperature Heat Sources— Part I: Theoretical Investigation," *ASME Journal of Solar Energy Engineering*, **125**(2), 1996.

APPENDIX A

SOLAR THERMAL AND PHOTOVOLTAIC COLLECTOR MANUFACTURING ACTIVITIES 2005*

OVERVIEW

U.S. manufacture of both solar thermal collector and photovoltaic (PV) cells and modules continued grew at a strong pace in 2005, despite the fact that prices for solar panels and PV cells and modules rose due to material cost increases. The solar industry was able to absorb most of the rising material costs because it had become more flexible in its production methods and supply arrangements over past years. It recovered from the nationwide economic downturn in 2003, and showed significant growth in 2004 and 2005.

SOLAR THERMAL COLLECTORS

Domestic shipments of solar thermal collectors rose 10.4 percent to 14.7 million square feet in 2005 (Table 1). There were 25 companies shipping solar collectors in 2005, one more than in 2004. Total shipments rose to 16 million square feet, a 13.7 percent increase over 2004. Exports surged 67.4 percent, while imports increased 22.1 percent (Table 2).

Low-temperature solar collectors represented 95 percent of total shipments. Medium-temperature collectors were responsible for more than 4 percent of total shipments (Table 3). High-temperature collectors represented less than 1 percent (0.7%). Included in the statistics were collectors shipped to Arizona Public Service's (APS) Saguaro Solar Trough Power Plant, the first concentrating solar power plant built in the U.S. Since 1988. The Saguaro Solar Trough Power Plant features more than 100,000 square feet of parabolic-trough shaped mirrors and stands more than 15 feet tall. It was built on a patch of desert in Red Rock, adjacent to APS' Saguaro Power Plant, about 30 miles north of Tucson. It has the capability of generating one megawatt of clean electrical power, enough electricity to meet the demands of about 200 homes (Figure 1).

In 2005, 71 percent of all collectors were produced in five states: New Jersey, California, Florida, Tennessee, and Arizona, with 63 percent of the total shipped from New Jersey and California alone. Twenty-eight percent of all collectors shipped were imported, mostly from Israel. More than 70 percent of all collectors were shipped to the top five domestic destinations: Florida, California, Arizona,

*Adapted from report released by the Energy Information Administration in August 2006 http://www.eia.doe.gov/cneaf/solar.renewables/page/solarreport/solar.html)

Table 1 Annual Solar Thermal Collector Domestic Shipments, 1996–2005

Year	Solar Thermal Collectors[a] (Thousand Square Feet)
1996	7,162
1997	7,759
1998	7,396
1999	8,046
2000	7,857
2001	10,349
2002	11,004
2003	10,926
2004	13,301
2005[p]	14,680
Total	**98,481**

[a] Total shipments minus export shipments.

[p] = Preliminary.

Note: Totals may not equal sum of components due to independent rounding. Total shipments include those made in or shipped to U.S. Territories.

Source: Energy Information Administration, Form EIA-63A, "Annual Solar Thermal Collector Manufacturers Survey."

New York, and Illinois. Florida and California accounted for 60 percent of total shipments (Table 4).

As indicated in Table 5, domestic shipments were sent to all 50 States within the U.S., plus the District of Columbia, Guam, Puerto Rico, and the U.S. Virgin Islands.

Exports experienced a record growth from 0.8 million square feet to 1.4 million square feet, mainly to Canada (36.37 percent), Brazil (20.97 percent), France (9.54 percent), and Mexico (8.14 percent) (Table 6). Fifty-eight percent of total shipments were sent directly to wholesale distributors, 33 percent to retail distributors, 4 percent to exporters, 4 percent to installers, and more than 1 percent to other end users (Table 7).

In general, the market was heavily dominated by low-temperature collectors for water heating applications (mainly swimming pool heating). Not surprisingly, the residential sector was the largest market for solar thermal collectors in 2005. Solar thermal collectors shipped to the residential sector in 2005 totaled 14.7 million square feet, nearly 92 percent of total shipments. The distant second-largest market for solar thermal collectors was the commercial sector, which accounted for only 1.2 million square feet, or about 7 percent of total shipments. The

Table 2   Annual Shipments of Solar Thermal Collectors, 1996–2005

Year	Number of Companies	Collector Shipments[a] (Thousand Square Feet)		
		Total[b]	Imports	Export
1996	28	7,616	1,930	454
1997	29	8,138	2,102	379
1998	28	7,756	2,206	360
1999	29	8,583	2,352	537
2000	26	8,354	2,201	496
2001	26	11,189	3,502	840
2002	27	11,663	3,068	659
2003	26	11,444	2,986	518
2004	24	14,114	3,723	813
2005[p]	25	16,041	4,546	1,361

[a]Includes imputation of shipment data to account for nonrespondents.

[b]Includes shipments of solar thermal collectors to the government, including some military, but excluding space applications.

[p] = Preliminary.

Note: Total shipments as reported by respondents include all domestic and export shipments and may include imported collectors that subsequently were shipped to domestic or foreign customers. *Source*: Energy Information Administration, Form EIA-63A, "Annual Solar Thermal Collector Manufacturers Survey."

largest end use for solar thermal collectors shipped in 2005 was for heating swimming pools, representing nearly 94 percent of the total shipments or 15 million square feet shipped. The distant second-largest end use for solar thermal collectors shipped in 2005 was domestic hot water systems, consuming 4 percent of the total shipments or 0.6 million square feet (Table 8).

The number of complete systems rose 72 percent to 51,265 systems in 2005. However, the value of complete systems increased 12 percent only (Table 9). This was mainly caused by more small systems being shipped in 2005 compared to fewer larger systems with almost the same value and total square feet in the prior year.

As in the previous years, the industry remained highly concentrated, with 92 percent of sales made by the 5 largest companies (Table 10). Employment increased more than 11 percent in 2005 (Table 11) to its second highest level over the past 10 years. A total of 22 companies were involved in the design of collectors or systems, 11 were involved in prototype collector development, and 11 were active in prototype system development (Table 12).

PHOTOVOLTAIC CELLS AND MODULES

The photovoltaic (PV) cell and module domestic shipments reached a record high of 134,465 peak kilowatts in 2005, a substantial 72 percent increase from

Table 3 Annual Shipments of Solar Thermal Collectors by Type, 1996–2005 (Thousand Square Feet)

| Year | Low Temperature | | Medium Temperature | | High Temperature |
	Total Shipments[a,b]	Average per Manufacturer	Total Shipments[a]	Average per Manufacturer	Total Shipments[a,c]
1996	6,821	487	785	41	10
1997	7,524	579	606	29	7
1998	7,292	607	443	23	21
1999	8,152	627	427	21	4
2000	7,948	723	400	25	5
2001	10,919	1,092	268	16	2
2002	11,126	856	535	31	2
2003	10,877	906	560	33	7
2004	13,608	1,512	506	30	0
2005[p]	15,224	1,522	702	41	115

[a] Includes imputation of shipment data to account for nonrespondents.

[b] Includes shipments of solar thermal collectors to the government, including some military, but excluding space applications.

[c] For high-temperature collectors, average annual shipments per manufacturer are not disclosed.

[p] = Preliminary.

Source: Energy Information Administration, Form EIA-63A, "Annual Solar Thermal Collector Manufacturers Survey."

Figure 1 APS Saguaro Solar Trough Power Plant. *Source*: Courtesy of Arizona Public Service (APS).

Table 4 Shipments of Solar Thermal Collectors
Ranked by Origin and Destination, 2005

Origin/Destination	2005 Shipments[p]	
	Thousand Square Feet	Percent of U.S. Total
Origin		
Top Five States	11,328	71
New Jersey	5,130	32
California	4,961	31
Florida	933	6
Tennessee	190	1
Arizona	114	1
Other Domestic	166	1
Imported	4,546	28
U.S. Total	**16,041**	**100**
Destination		
Top Five States	11,299	70
Florida	5,408	34
California	4,137	26
Arizona	794	5
New York	499	3
Illinois	461	3
Other Domestic	3,381	21
Exported	1,361	8
U.S. Total	**16,041**	**100**

[*] = Less than 0.5 percent.
[p] = Preliminary.
Note: Totals may not equal sum of components due to independent rounding. U.S. total includes territories.
Source: Energy Information Administration, Form EIA-63A, "Annual Solar Thermal Collector Manufacturers Survey."

the 2004 record of 78,346 peak kilowatts, and was an increase of more than 176 percent from the 2003 level (Table 13). Rising electricity prices during 2004–2005 years increased demand for PV products, which spawned new PV technology and business opportunities during 2005.

Total shipments of PV cells and modules rose to 226,916 peak kilowatts in 2005, a 25 percent increase over the 2004. Module shipments increased 43 percent to 204,996 peak kilowatts, but cell shipments decreased to 21,920 peak kilowatts from 37,842 peak kilowatts in 2004 (Table 14). This suggested a potential shift in manufacturer focus of offering unique PV modules to meet the strong demand of their customers, likely caused, in part, by higher energy prices.

Prior to 2005, the number of active companies shipping PV cells and modules had remained steady, averaging 20 over the past two decades. In 2005, however,

Table 5 Shipments of Solar Thermal Collectors by Destination, 2005 (Square Feet)

Destination	Shipments[p]
Alabama	51,306
Alaska	324
Arizona	794,477
Arkansas	22,104
California	4,136,510
Colorado	62,931
Connecticut	327,876
Delaware	676
District of Columbia	350
Florida	5,407,966
Georgia	47,241
Guam	328
Hawaii	363,282
Idaho	15,782
Illinois	461,368
Indiana	50,341
Iowa	16,268
Kansas	18,437
Kentucky	15,961
Louisiana	23,401
Maine	28,005
Maryland	25,007
Massachusetts	73,253
Michigan	237,464
Minnesota	28,903
Mississippi	1,924
Missouri	16,939
Montana	530
Nebraska	16,351
Nevada	284,422
New Hampshire	23,420
New Jersey	424,670
New Mexico	15,804
New York	498,918
North Carolina	142,409
North Dakota	3,208
Ohio	34,663
Oklahoma	14,970
Oregon	269,251
Pennsylvania	233,797
Puerto Rico	116,737
Rhode Island	16,227

Table 5 (*continued*)

Destination	Shipments[p]
South Carolina	3,191
South Dakota	509
Tennessee	1,811
Texas	47,948
Utah	2,677
Vermont	12,938
Virgin Islands of the U.S.	4,086
Virginia	221,762
Washington	16,265
West Virginia	13,241
Wisconsin	31,148
Wyoming	485
Shipments to United States/Territories	14,679,862
Exports	1,361,116
Total Shipments	**16,040,978**

[p] = Preliminary.
Source: Energy Information Administration, Form EIA-63A, "Annual Solar Thermal Collector Manufacturers Survey."

the number of active companies surged to 29, compared to just 19 in 2004. Imports jumped to 90,981 peak kilowatts in 2005 from 47,703 peak kilowatts in 2004, an increase of 91 percent. The main contributors to the increase were American subsidiaries of Japanese companies who were principally importing cells. In contrast, exports dropped to 92,451 peak kilowatts in 2005 from 102,770 peak kilowatts in the previous year, a decrease of 10 percent (Table 15).

Shipments to wholesale distributors, the largest business category, increased more than 22 percent from 106,400 peak kilowatts in 2004 to 130,086 peak kilowatts in 2005. Shipments to the second-largest business category, installers, surged 94 percent to 67,437 peak kilowatts in 2005 (Table 16).

Although the market share of crystalline silicon cells and modules declined to 76 percent from 88 percent in 2004, it was still the dominant type of solar cell. Within that category, single-crystal shipments fell to 71,901 peak kilowatts, or slightly less than 32 percent of total shipments in 2005, compared to 94,899 peak kilowatts in 2004. In contrast, cast and ribbon silicon shipments rose to 101,065 peak kilowatts in 2005, or close to 45 percent of total shipments, compared to 64,239 peak kilowatts in 2004. Cast and ribbon became the predominant PV technology. Fueled by the rapidly growing market, and the continuing tight silicon supply–thin film technology uses less silicon per unit of electrical output than does crystalline silicon technology–shipments of the small thin-film market more than doubled to 53,826 peak kilowatts in 2005, compared to 21,978 peak

Table 6 Distribution of U.S. Solar Thermal Collector Exports by Country, 2005

Country	U.S. Export Shipments (Square Feet)^p	Percent of U.S. Export
Africa		
Reunion	1,584	0.12
Total	**1,584**	**0.12**
Europe		
Austria	14,950	1.10
Belgium	12,888	0.95
Czech Republic	11,775	0.87
Federal Republic of Germany	75,000	5.51
France	129,801	9.54
Italy	10,891	0.80
Spain	52,198	3.83
Sweden	49,172	3.61
Switzerland	2,880	0.21
Total	**359,555**	**26.42**
North & Central America		
Antigua and Barbuda	2,128	0.16
Bahamas	2,471	0.18
Bermuda	971	0.07
Canada	495,048	36.37
Cayman Islands	380	0.03
Costa Rica	4,305	0.32
Dominican Republic	1,426	0.10
Guatemala	6,598	0.48
Jamaica	125	0.01
Mexico	110,740	8.14
Netherlands Antilles	126	0.01
Trinidad and Tobago	1,200	0.09
Turks and Caicos Islands	2,950	0.22
Total	**628,468**	**46.17**
Oceania & Australia		
Australia	71,989	5.29
New Zealand	13,989	1.03
Total	**85,978**	**6.32**
South America		
Brazil	285,451	20.97
Ecuador	80	0.01
Total	**285,531**	**20.98**
Total	**1,361,116**	**100.00**

^{p} = Preliminary.
* = Less than 0.01 percent.
Note: Totals may not equal sum of components due to independent rounding.
Source: Energy Information Administration, Form EIA-63A, "Annual Solar Thermal Collector Manufacturers Survey."

Table 7 Distribution of Solar Thermal Collector
Shipments, 2004 and 2005

Recipient	Shipments (Thousand Square Feet)	
	2004	2005^P
Wholesale Distribution	8,248	9,248
Retail Distributors	5,092	5,342
Exporters	253	571
Installers	398	633
End Users and Other^a	124	248
Total	**14,114**	**16,041**

^{a}Other includes minimal shipments not explained on form EIA-63A.
^{P} = Preliminary.
Note: Totals may not equal sum of components due to independent
rounding. Total includes U.S. territories.
Source: Energy Information Administration, Form EIA-63A, "Annual
Solar Thermal Collector Manufacturers Survey."

kilowatts in 2004. Thin-film accounted for nearly one-fourth of the PV market
(Table 17).

The total value of PV cell and module shipments grew nearly 40 percent to
$701.7 million in 2005 (Table 18).

Among the market sectors, the commercial sector remained the largest sector
for PV shipments, followed by the residential and industrial sectors. Commercial
sector shipments totaled 89,459 peak kilowatts and grew at a rate of 20 percent from 2004 to 2005. The residential sector totaled 75,040 peak kilowatts in
2005, increasing more than 39 percent over the previous year. Electricity generation, which consists of both grid-interactive (those connected to electric power
grid) and remote applications (those not connected), continued to be the predominant end use for PV cells and modules. In 2005, electric generation was
about 85 percent of the total shipments, and was 31 percent more than in 2004
(Table 19).

Nearly 78 percent of PV exports were modules during 2005 (Table 20). Shipments to Europe represented more than 72 percent of total U.S. exports, with
Germany being responsible for slightly over 53 percent of the total. Although
the Netherlands continued as the second-largest U.S. export market, exports to
the Netherlands declined from 28,744 peak kilowatts in 2004 to 11,997 peak
kilowatts in 2005 (Table 21).

Shipments of complete PV systems surged 118 percent from 16,990 systems
in 2004 to 37,115 systems in 2005. While the total value of completed systems
increased 9 percent to $43.0 million, total peak kilowatts dropped from 8,110
in 2004 to 6,583 in 2005. These statistics reflected the evolution of thin-film

Table 8 Shipments of Solar Thermal Collectors by Market Sector, End Use, and Type, 2004 and 2005 (Thousand Square Feet)

	Low Temperature Liquid/Air		Medium Temperature Liquid				High Temperature		
	Metallic and Nonmetallic	Air	ICS/Thermo-siphon	Flat-Plate (Pumped)	Evacuated Tube	Concentrator	Parabolic Dish/Trough	2005 Total^p	2004 Total
Market Sector									
Residential	14,045	3	151	479	3	0	0	14,681	12,864
Commercial	1,099	0	12	46	*	0	2	1,160	1,178
Industrial	30	0	1	0	0	0	0	31	70
Utility	0	0	0	0	0	0	114	114	0
Other^a	50	0	*	6	0	0	0	56	3
Total	**15,224**	**3**	**165**	**530**	**3**	**0**	**115**	**16,041**	**14,114**
End use									
Pool Heating	15,022	0	0	20	*	0	0	15,041	13,634
Hot Water	12	0	165	461	2	0	0	640	452
Space Heating	190	3	0	34	1	0	0	228	13
Space Cooling	0	0	0	0	0	0	2	2	0
Combined Space and Water Heating	0	0	0	16	0	0	0	16	16
Process Heating	0	0	0	0	0	0	0	0	0
Electricty Generation	0	0	0	0	0	0	114	114	0
Other^b	0	0	0	0	0	0	0	0	0
Total	**15,224**	**3**	**165**	**530**	**3**	**0**	**115**	**16,041**	**14,114**

^{a}Other market sector includes shipments of solar thermal collectors to sectors such as government, including the military but excluding space applications.
^{b}Other end use includes shipments of solar thermal collectors for other uses such as cooking, water pumping, water purification, desalination, distillation, etc.
* = Less than 500 square feet.
ICS = Integral Collector Storage.
^{p} = Preliminary.
Note: Totals may not equal sum of components due to independent rounding.
Source: Energy Information Administration, Form EIA-63A, "Annual Solar Thermal Collector Manufacturers Survey."

284

Table 9   Shipments of Complete Solar Thermal Collector Systems, 2003 and 2004

Shipment Information	2004	2005^P
Complete Collector Systems		
Shipped	29,769	51,265
Thousand Square Feet	5,560	5,748
Percent of Total Shipments	39	36
Number of Companies	18	18
Value of Systems (Thousand Dollars)	18,293	20,402

^{P} = Preliminary.
Source: Energy Information Administration, Form EIA-63A, "Annual Solar Thermal Collector Manufacturers Survey."

Table 10   Percent of Solar Thermal Collectors Shipments by 10 Largest Companies, 1996–2005

Year	Company Rank	Shipments (Thousand Square Feet)	Percent of Total Shipments
1996	1–5	6,452	85
	6–10	910	12
1997	1–5	7,183	88
	6–10	731	9
1998	1–5	6,938	89
	6–10	613	8
1999	1–5	7,813	91
	6–10	563	7
2000	1–5	7,521	90
	6–10	567	7
2001	1–5	10,732	96
	6–10	325	3
2002	1–5	10,755	92
	6–10	670	6
2003	1–5	10,485	92
	6–10	700	6
2004	1–5	13,291	94
	6–10	664	5
2005^P	1–5	14,801	92
	6–10	934	6

^{P} = Preliminary.*Note*: Totals may not equal sum of components due to independent rounding.
Source: Energy Information Administration, Form EIA-63A, "Annual Solar Thermal Collector Manufacturers Survey."

Table 11
Employment in the
Solar Thermal Collector
Industry, 1996–2005

Year	Person Years
1996	239
1997	184
1998	207
1999	289
2000	284
2001	256
2002	356
2003	287
2004	317
2005^p	353

p = Preliminary.
Source: Energy Information Administration, Form EIA-63A, "Annual Solar Thermal Collector Manufacturers Survey."

Table 12   Companies Involved in Solar Thermal Collector Activities by Type, 2004 and 2005

Type of Activity	2004	2005^p
Collector or System Design	19	22
Prototype Collector Development	10	11
Prototype System Development	8	11
Wholesale Distribution	22	23
Retail Distribution	11	11
Installation	8	9
Noncollector System Component Manufacture	11	10

p = Preliminary.
Source: Energy Information Administration, Form EIA-63A, "Annual Solar Thermal Collector Manufacturers Survey."

technology, as the systems shipped in 2005 were smaller, more flexible, and lighter-weight compared to conventional PV systems (Table 22).

Employment in the PV manufacturing industry increased more than 6 percent, from 2,916 person-years in 2004 to 3,108 person-years in 2005 (Table 23).

Table 13 Annual Photovolataic Domestic Shipments, 1996–2005

Year	Photovoltaic Cells and Modules[a] (Peak Kilowatts)
1996	13,016
1997	12,561
1998	15,069
1999	21,225
2000	19,838
2001	36,310
2002	45,313
2003	48,664
2004	78,346
2005[p]	134,465
Total	**424,807**

Total shipments minus export shipments.
[p] = Preliminary.
Note: Totals may not equal sum of components due to independent rounding. Total shipments include those made in or shipped to U.S. Territories.
Sources: Energy Information Administration, Form EIA-63B, "Annual Photovoltaic Module/Cell Manufacturers Survey."

Table 14 Annual Shipments of Photovolataic Cells and Modules, 2003–2005 (Peak Kilowatts)

Item	2003	2004	2005[p]
Cells	29,295	37,842	21,920
Modules	80,062	143,274	204,996
Total	**109,357**	**181,116**	**226,916**

[p] = Preliminary.
Sources: Energy Information Administration, Form EIA-63B, "Annual Photovoltaic Module/Cell Manufacturers Survey."

Table 24 shows that of the companies involved in PV-related activities, twelve were involved in cell manufacturing and twenty-three in module or systems design. Eighteen were involved in prototype module development and nine in prototype systems development. Nineteen companies were active in wholesale distribution, seven in retail distribution, and seven were involved in installation.

Table 15   Annual Shipments of Photovolataic Cells and Modules, 1996–2005

| Year | Number of Companies | Photovoltaic Cell and Modules Shipments[a] (Peak Kilowatts) | | |
		Total	Imports	Exports
1996	25	35,464	1,864	22,448
1997	21	46,354	1,853	33,793
1998	21	50,562	1,931	35,493
1999	19	76,787	4,784	55,562
2000	21	88,221	8,821	68,382
2001	19	97,666	10,204	61,356
2002	19	112,090	7,297	66,778
2003	20	109,357	9,731	60,693
2004	19	181,116	47,703	102,770
2005[p]	29	226,916	90,981	92,451

[a] Does not include shipments of cells and modules for space/satellite applications.

[p] = Preliminary.

Note: Total shipments as reported by respondents include all domestic and export shipments and may include imported cells and modules that subsequently were shipped to domestic or foreign customers.

Source: Energy Information Administration, Form EIA-63B, "Annual Photovoltaic Module/Cell Manufacturers Survey."

Table 16   Distribution of Photovoltaic Cells and Modules, 2003–2005

| Recipient | Shipments (Peak Kilowatts) | | |
	2003	2004	2005[p]
Wholesale Distributers	65,477	106,400	130,086
Retail Distributers	6,624	5,140	2,362
Exporters	7,600	2,354	1,088
Installers	11,733	34,779	67,437
End-Users	8,286	1,029	3,142
Module Manufacturers	8,738	11,868	15,347
Other[a]	899	19,546	7,455
Total	**109,357**	**181,116**	**226,916**

[a] Other includes categories not identified by reporting companies.

[p] = Preliminary.

Note: Totals may not equal sum of components due to independent rounding.

Source: Energy Information Administration, Form EIA-63B, "Annual Photovoltaic Module/Cell Manufacturers Survey."

Table 17   Photovolataic Cell and Module Shipments by Type, 2003–2005

Type	Shipments (Peak kilowatts)			Percent of Total		
	2003	2004	2005^p	2003	2004	2005^p
Crystalline Silicon						
Single-Crystal	59,379	94,899	71,901	54	52	32
Cast and Ribbon	38,561	64,239	101,065	35	35	45
Subtotal	97,940	159,138	172,965	90	88	76
Thin-Film	10,966	21,978	53,826	10	12	24
Concentrator	452	0	125	*	0	*
Other[a]	0	0	0	0	0	0
Total	**109,357**	**181,116**	**226,916**	**100**	**100**	**100**

[a]Includes categories not identified by reporting companies.
* = Less than 0.5 percent.
^{p} = Preliminary.
Note: Data do not include shipments of cells and modules for space/satellite applications. Totals may not equal sum of components due to independent rounding.
Source: Energy Information Administration, Form EIA-63B, "Annual Photovoltaic Module/Cell Manufacturers Survey."

Table 18   Photovoltaic Cell and Module Shipment Values by Type, 2004 and 2005

	2004			2005^p		
	Value (Thousand Dollars)	Average Price (Dollars per Peak Watt)		Value (Thousand Dollars)	Average Price (Dollars per Peak Watt)	
		Modules	Cells		Modules	Cells
Crystalline Sillicon						
Single-Crystal	253,558	3.09	1.94	227,751	3.48	2.2
Cast and Ribbon	188,371	3	1.76	318,690	3.2	2.02
Subtotal	441,930	3.04	1.92	546,440	3.3	2.17
Thin-Film Silicon	W	W	W	W	W	W
Concentrator Silicon	W	W	W	W	W	W
Other[a]	0	–	–	0	–	–
Total	**501,739**	**2.99**	**1.92**	**701,718**	**3.19**	**2.17**

^{W} = Data withheld to avoid disclosure of proprietary company data.
$\hat{}${a}$Includes categories not identified by reporting companies. $^-$ = Does not apply.
^{p} = Preliminary.
 Note: Data do not include shipments of cells and modules for space/satellite applications. Totals may not equal sum of components due to independent rounding.
Source: Energy Information Administration, Form EIA-63B, "Annual Photovoltaic Module/Cell Manufacturers Survey."

Table 19   Shipments of Photovoltaic Cells and Modules by Market Sector, End Use, and Type, 2004 and 2005 (Peak Kilowatts)

Sector and End Use	Crystalline Silicon[a]	Thini-Film Silicon	Concentrator Silicon	Other	2005 Total[p]	2004 Total
Market						
Industrial	21,674	525	0	0	22,199	30,493
Residential	70,986	4,029	25	0	75,040	53,928
Commercial	61,084	28,349	25	0	89,459	74,509
Transportation	1,621	0	0	0	1,621	1,380
Utility	68	0	75	0	143	3,233
Government[b]	8,034	20,649	0	0	28,683	3,257
Other[c]	9,498	274	0	0	9,772	14,316
Total	**172,965**	**53,826**	**125**	**0**	**226,916**	**181,116**
End Use						
Electricity Generation						
Grid	126,157	42,217	100	0	168,474	129,265
Interactive Remote	23,589	1,344	25	0	24,958	18,371
Communication	8,507	159	0	0	8,666	11,348
Consumer Goods	5,511	276	0	0	5,787	6,444
Transportation	2,159	0	0	0	2,159	1,380
Water Pumping	1,273	70	0	0	1,343	1,322
Cells/Modules to OEM[d]	2,008	9,669	0	0	11,677	6,452
Health						341
Other[e]	3,762	91	0	0	3,853	6,193
Total	**172,965**	**53,826**	**125**	**0**	**226,916**	**181,116**

[a] Includes single-crystal and cast and ribbon types.

[b] Includes Federal, State, local governments, excluding military.

[c] Other includes shipments that are manufactured for private contractors for research.

[d] Original equipment manufacturer.

[e] Other includes shipments of photovoltaic cells and modules for other uses, such as cooking food, desalinization, distillation, etc.

[p] = Preliminary.

Note: Totals may not equal sum of components due to independent rounding.

Source: Energy Information Administration, Form EIA-63B, "Annual Photovoltaic Module/Cell Manufacturers Survey."

Table 20 Export Shipments of Photovoltaic Cells and Modules by Type, 2004 and 2005 (Peak Kilowatts)

Item	Crystalline 2004	2005^p	Thin-Film Silicon 2004	2005^p	Concentrator Silicon 2004	2005^p	Total 2004	2005^p
	2004	**2005^p**	**2004**	**2005^p**	**2004**	**2005^p**	**2004**	**2005^p**
Cells	36,492	20,434	0	0	0	0	36,492	20,434
Modules	52,938	39,992	13,341	32,000	0	25	66,278	72,017
Totals	**89,430**	**60,426**	**13,341**	**32,000**	**0**	**25**	**102,770**	**92,451**

^{p} = Preliminary. *Notes*: Totals may not equal sum of components due to independent rounding.
Source: Energy Information Administration, Form EIA-63B, "Annual Photovoltaic Module/Cell Manufacturers Survey."

Table 21 Destination of U.S. Photovolataic Cell and Module Export Shipments by Country, 2005

Country	Peak Kilowatts^p	Percent of U.S. Exports
Africa		
Angola	0.3	*
Egypt	232.3	0.3
Gambia	1.3	*
Kenya	84.0	0.1
Nigeria	76.7	0.1
South Africa	548.5	0.6
Total	**943.1**	**1.0**
Asia		
China	1,938.7	2.1
Hong Kong	2,935.1	3.2
India	1,480.2	1.6
Israel	14.0	*
Japan	1,085.2	1.2
Malaysia	1.9	*
Nepal	93.0	0.1
North Korea	78.4	0.1
Oman	64.0	0.1
Pakistan	64.3	0.1
Philippines	37.0	*
Saudi Arabia	1.0	*
Singapore	8,560.2	9.3
South Korea	575.3	0.6
SriLanka	12.9	*

Table 21   (*Continued*)

Country	Peak Kilowatts^p	Percent of U.S. Exports
Taiwan	114.3	0.1
Thailand	101.0	0.1
United Arab Emirates	1.0	*
Vietnam	3.0	*
Total	**17,160.5**	**18.6**
Europe		
Austria	587.0	0.6
Belgium	4.0	*
Denmark	56.0	0.1
Federal Republic of Germany	49,249.9	53.3
Finland	20.0	*
France	43.0	*
Italy	673.1	0.7
Kazakhstan	1.1	*
Luxembourg	925.0	1.0
Netherlands	11,996.7	13.0
Norway	0.2	*
Poland	1.0	*
Portugal	1,902.0	2.1
Russia	17.0	*
Slovakia	90.0	0.1
Spain	706.4	0.8
Sweden	0.2	*
Switzerland	183.8	0.2
Turkey	1.6	*
United Kingdom	555.2	0.6
Uzbekistan	1.0	*
Total	**67,014.2**	**72.5**
North & Central America		
Antigua and Barbuda	1.6	*
Bermuda	1.0	*
Canada	3,226.5	3.5
Costa Rica	342.6	0.4
Dominican Republic	64.4	0.1
Guadeloupe	271.6	0.3
Guatemala	16.2	*
Haiti	53.7	0.1
Honduras	32.5	*
Martinique	4.6	*
Mexico	1,073.7	1.2
Netherlands Antilles	14.3	*
Nicaragua	0.8	*

Table 21   (*Continued*)

Country	Peak Kilowatts^p	Percent of U.S. Exports
Panama	56.2	0.1
Trinidad and Tobago	1.0	*
Total	**5,160.7**	**5.6**
Oceania & Australia		
Australia	1,006.0	1.1
New Zealand	66.3	0.1
Total	**1,072.3**	**1.2**
South America		
Argentina	120.3	0.1
Bolivia	33.7	*
Brazil	461.1	0.5
Chile	39.7	*
Colombia	55.0	0.1
Ecuador	2.5	*
Guyana	16.5	*
Peru	355.2	0.4
Uruguay	1.2	*
Venezuela	14.9	*
Total	**1,100.1**	**1.2**
Total U.S. Export Export	**92,450.9**	**100.0**

^{p} = Preliminary.
* = Value less than 0.05 percent.*Note*: Totals may not equal sum of components due to independent rounding.
Source: Energy Information Administration, Form EIA-63B, "Annual Photovoltaic Module/Cell Manufacturers Survey."

Table 22   Shipments of Complete Photovoltaic Systems, 2003–2005

Shipment Information	2003	2004	2005^p
Complete Photovoltaic Module System Shipped	5,525	16,990	37,115
Peak Kilowatts	9,545	8,110	6,583
Percentage of Total Module Shipments	12	6	3
Value of Systems (Thousand Dollars)	50,412	39,459	43,029

^{p} = Preliminary.
Source: Energy Information Administration, Form EIA-63B, "Annual Photovoltaic Module/Cell Manufacturers Survey."

Table 23   Employment in the Photovoltaic Manufacturing Industry, 1996–2005

Year	Number of Companies	Number of Person Years
1996	25	1,280
1997	21	1,736
1998	21	1,988
1999	19	2,013
2000	21	1,913
2001	19	2,666
2002	19	2,696
2003	20	2,590
2004	19	2,916
2005[p]	29	3,108

[p] = Preliminary.
Source: Energy Information Administration, Form EIA-63B, "Annual Photovoltaic Module/Cell Manufacturers Survey."

Table 24   Number of Companies Involved in Photovoltaic-Related Activities, 2004 and 2005

Type of Activity	Number of Companies	
	2004	2005[p]
Cell Manufacturing	12	12
Module or Systems Design	18	23
Prototype Module Development	13	18
Prototype Systems Development	9	9
Wholesale Distribution	16	19
Retail Distribution	10	7
Installation	6	7
Noncollector System Component Manufacturing	3	3

[p] = Preliminary.
Source: Energy Information Administration, Form EIA-63B, "Annual Photovoltaic Module/Cell Manufacturers Survey."

APPENDIX B

Survey of Geothermal Heat Pump Shipments, 1990–2004*

Based on the Energy Information Administration, Form EIA-902, "Annual Geothermal Heat Pump Manufacturers Survey," manufacturers shipped 43,806 geothermal heat pumps in 2004, a 20 percent increase over the 2003 total of 36,439.

The proportion of geothermal heat pumps shipped to each Census Region in 2004 was as follows: the South (33 percent), the Midwest (33 percent), the Northeast (18 percent), and the West (8 percent). The proportion of geothermal heat pumps exported was 7 percent.

Analysis conducted by the Oregon Institute of Technology, Geo-Heat Center, indicated that geothermal heat pumps consumed almost 29 trillion Btu of geothermal energy in 2004 and direct uses, such as crop drying, consumed 9 trillion Btu of geothermal energy (Table 1).

Table 1 Geothermal Direct Use of Energy and Heat Pumps, 1990–2004 (Quadrillion Btu)

Year	Direct Use	Heat Pumps	Total
1990	0.0048	0.0054	0.0102
1991	0.0050	0.0060	0.0110
1992	0.0051	0.0067	0.0118
1993	0.0053	0.0072	0.0125
1994	0.0056	0.0076	0.0132
1995	0.0058	0.0083	0.0141
1996	0.0059	0.0093	0.0152
1997	0.0061	0.0101	0.0162
1998	0.0063	0.0115	0.0178
1999	0.0079	0.0114	0.0193
2000	0.0084	0.0122	0.0206
2001	0.0090	0.0135	0.0225
2002	0.0090	0.0147	0.0237
2003	0.0086	0.0274	0.0360
2004	0.0090	0.0289	0.0379

Note: Direct use includes applications such as: district heating, aquaculture pond and raceway heating, greenhouse heating and agricultural drying.

Source: John Lund, Oregon Institute of Technology, Geo-Heat Center (Klamath Falls, Oregon, March 2005).

*Adapted from report released by the Energy Information Administration in March 2006(http://www.eia.doe.gov/cneaf/solar.renewables/page/ghpssurvey.html)

Index

A

Absorption chillers, 142–143
Acid rain, emissions (improvements), 214f
Activation polarization, 73–74
AFC. *See* Alkaline fuel cell
Agglomeration. *See* Preconversion technology
Air-based systems, usage, 42–43
Aircraft (aero-derivative) gas turbines, 136–137
Air pollution, regulations. *See* Cogeneration
Air Quality Control Regions (AQCR), 150
Air reactors, 254
Alkaline electrolyzer, water (electrolysis), 171f
Alkaline fuel cell (AFC), 60, 91
Alkaline water electrolyzer, 172
Alternative Energy Research, ammonia-water combined cycle research, 273
American Council for an Energy Efficient Economy (ACEEE) report, 269
Amine scrubbing, 255
Ammonia/water combined cycle flow diagram, 273f
Annual solar thermal collector domestic shipments, 276t
Anthracite, characteristics, 209
Arizona Power Supply (APS) Saguaro Solar Trough Power Plant, 278f
Aromatic sulfonic acid polymers, synthesis/characterization, 172
Artificial efficiency, 148
ASHRAE Standard Number, 96–80, 38
ASTM, E905 standard, 38
Atmospheric absorption spectrum, 212f
Attainment rate, 154
Auxiliary power unit (APU), 60

B

Bacteria, fermentation, 176
Bag filters, 215
   pulse cleaning/back-flushing, 241
   usage, 239–241
Bag house, 240f
Balance of Plant (BOP) equipment, 143
Base-load system, 145
Best available control technology (BACT), 151–153
BFB. *See* Bubbling fluidized bed
Binary cycle conversion, 113–115
   schematic, 114f
   variations, 114
Binary power plants, 109
Biological hydrogen production, concept, 176f
Biomass
   components, 170
   direct gasification, 169–170
   usage. *See* Hydrogen production
Biophotolysis, 176
Bituminous coal, characteristics, 209
Boronhydride complexes, usage, 188
Bottoming cycles, 133–134
Bubbling fluidized bed (BFB), 224, 225–226
   combustor, 225f
Butler-Volmer equation, usage, 73

C

CAA. *See* Clean Air Act
CAAA. *See* Clean Air Act Amendments
Capital recovery factor (CRF), 4
Carbon capture/sequestration, cost, 260t
Carbon dioxide (CO_2)
   capture, 252–257
      amine scrubber, usage, 256f
      technologies, techno-economic comparison, 260t
   combustion, 253–255
   control, 215
   cost implications, 259–261
   emissions, reduction, 221f
      means, 251f
   extraction. *See* Postcombustion CO_2 extraction

Carbon dioxide (CO_2) *(Continued)*
  industrial utilization, 259
  membrane scrubbing, 256
  pipeline transportation, 257
  pollutant, 212–213
  precombustion, 252
  production, 251–261
  separation
    cost, 256–257
    cryogenic distillation, usage, 256
    sorbent/solvents, usage, 255
  sequestration, 257–259
    cost, 260–261
    options, 258f
  ship transportation, 257
  transportation, 257
  underground sequestration, 258
  undersea sequestration, 258–259
  utilization, 257–259
Carbon sink management, 259
Carnot cycle, 222
Catalyst/electrolyte poisoning/degradation, 77
Catalyst layer, morphology changes/loss, 77
Central receiver power plant, schematic
    diagram, 37f
CFB. *See* Circulating fluidized bed
CFR. *See* Code of Federal Regulations
Chemical hydrogen storage, 188–189
Chemical looping combustion, 254f
  usage, 253–255
CHP. *See* Combined heat and power
Circulating fluidized bed (CFB), 224,
    226–228
  combustor, 227f
Clean Air Act Amendments (CAAA), 150
Clean Air Act (CAA) of, 1963, 150, 213
Clean Air Act of 1956 (Great Britain), 213
Clean Air Interstate Rule, 3
Clean Air Mercury Rule, 3, 213
Clean Water Act of, 1977, 153
Coal
  characteristics, 208–209
  cleaner energy, 214–215
    motivation, 213–214
  clean power generation
    introduction, 207–215
    references, 261–265
  conversion, 219–239

  impurities, 209
  pollutants, 209–213
Coal-fired IGCC plants, examples, 236
Coal-fired power plants
  emissions, improvements, 219t
  pollution reduction, 209t
Coal gasification, 169
Coallike fuel, properties, 210f
Code of Federal Regulations (CFR), gas
    turbine emission level requirements, 152
COE. *See* Cost of electricity
Cogeneration
  air pollution regulations, 150–151
  alternatives, economic comparisons, 10f
  back-up rates, problems, 132
  bottoming-cycle system, schematic, 133f
  capital investment/fuel, expense, 131
  considerations, 159–161
  constraints, 131–132
  conventional ownership/operation, 161
  cycles, 132–134
    combination, 134–135
  dynamic power/thermal matching, 145–146
  economic evaluations, 154–159
  economic merit, 157–159
  electrical equipment, usage, 139–140
  electrical/thermal loads, matching,
    144–145
  electricity sales, revenue restrictions, 132
  environmental concerns, 132
  equipment/components, description,
    139–143
  equipment specific regulations, 151–153
  facility, thermal/electric loads, 160
  federal regulations, 147–149
  final economic evaluation, template, 158t
  heat-recovery equipment, 140–142
  history, 130–131
  impact, 8–10
  introduction, 130–132
  ownership/financial arrangements, 159–163
  ownership/financing structures,
    characteristics (summary), 159t
  packaged systems, 146–147
  partnership arrangements, 161–162
  permits/certificates, 154
  power plant, operating modes, 145
  projects, financing options, 160
  references, 163–164
  regulatory considerations, 147–154
  solid waste disposal, 153–154

systems, 132–135
  applications, 135
  economic evaluations, actions, 155
  operating costs, 155–156
  proposal, operating costs, 157
  technical design issues, 143–147
  third-party ownership, 162
  topping-cycle system, schematic, 132f
  water quality, 153–154
Cogenerator, electric utility legal obligations, 149
Collectors
  carrier efficiency, 54
  performance, 32f
  plate temperature, measurement (difficulty), 31–32
  testing, 37–39
Column froth flotation coal cleaning, 218f
Combined-cycle cogeneration system, schematic, 134f
Combined heat and power (CHP) system, 268–269
  interconnection, 270
Combustion, T requirements, 220
Combustor designs, variety, 137–138
Complex hydrides
  hydrogen storage capacities, 187t, 188t
  usage potential, 186–188
Compound curvature concentrators, 35–37
Concentrating collectors, 33–37
Concentrating tables, usage, 216
Concentrators, optical efficiency, 35
Concentrator till factors, 29t
Conduction-dominated resources, 103
Convection suppression method, 45–46
Convention on Long-Range Transboundary Air Pollution, 213
Corrosion, impact, 77–78
Cost of electricity (COE), 119. *See also* Levelized cost of electricity
Cryogenic distillation, usage. *See* Carbon dioxide
Crystalline silicon cells/modules, market share (decline), 281–283
Crystalline solar cells, efficiencies, 56
Cyclone
  combustion, 221, 231–232
  combustor, 232f
  usage, 216–217

D

Daily solar flux conversions, 27–28
Degassing units, filtered steam (expansion), 111
Dense-media vessels, usage, 216, 217
Department of Energy Solid State Energy Conversion Alliance (SECA), 85
Depth filtration, 239–240
  caking, 240f
Deutsch-Anderson equation, 242
Diffuse radiation, values, 27
Diffusion driven desalination (DDD) method, 270–271
  process, flow diagram, 271f
Diffusion media (DM), 64
Direct-fired units, 142–143
Direct-gain passive heating system, 44f
Direct methanol fuel cell (DMFC), 62, 83
  cathode stoichiometry, minimum (determination), 84
  external humidification, requirement (absence), 83–84
  performance, transport-related issues, 84–85
  technical issues, 83–85
Direct steam conversion, 110–112
  schematic, 110f
Distributed power applications, 62
Double-tank indirect solar water-heating system, 40f
Dry scrubbing, usage, 245–246
Dry sorbent injection, usage, 245–246
Duct burner, 140
Dust filtration systems, types, 239–241
Dynamic power/thermal matching. *See* Cogeneration

E

Earth
  declination/hour angle, 14–16
  motion, 14f
  quantification, 15
ECBM. *See* Enhanced coal bed methane
ECSA. *See* Electrochemical active surface area
Electrical power generation, abandonment (factors), 131
Electrical/thermal loads
  example, 144f
  matching. *See* Cogeneration

Electricity, on-site cogeneration (fuel cell usage), 193
Electricity cost
  calculations, 5f
  operation hours, contrast, 7f
Electric power generation, geothermal usage, 109–116
Electrochemical active surface area (ECSA), 77
Electrochemical combustor, 191
Electrode reaction, stoichiometric ratio, 71
Electrolysis, 171–172
Electrolyte, 192
  loss, 77
Electrolyte/electrode assembly, usage, 64
Electrons, physical transport, 193
Electrostatic precipitators (ESPs), 215, 241–242
  advantages, 242
  operation, 242f
Elemental mercury, removal, 250
Endothermic gasification reactions, energy (providing), 233
Energy, geothermal direct use, 295t
Energy Policy Act (EPAct) of 2005, 149–150
Energy production (worldwide), 208f
Enhanced coal bed methane (ECBM) extraction, 258
Enhanced oil recovery (EOR), 257, 258
Equation of time (EoT), calculation, 16
ESPs. See Electrostatic precipitators
European Integrated Hydrogen Project specifications, 183
Extraterrestrial solar flux, 22–25

F

Facultative autotrophs, 219
Faraday effect, 140
Fast bed, 227
FBC. See Fluidized bed combustion
Federal air quality regulations, flowchart, 153f
Federal Energy Regulatory Commission (FERC), 147, 149
Federal Water Pollution Control Act of, 1956, 153
FGD. See Flue gas desulphurization
First-cut energy/cost analyses, 156
Fixed-bed gasifier, gas velocity, 233
Fixed charge rate (FCR), 4

Fixed circular trough, tracking absorber (inclusion), 34f
Flashed steam conversion, 112–113
  schematic, 112f
Flashing, 183
Flash power conversion, inefficiency, 113–114
Flat-plate collectors
  energy production, 31
  usage, 28–33
Flow field design, 81
Flue gas desulphurization (FGD), 215, 226, 244
  process. See Wet FGD process
  limed-based sorbent, usage, 250
Fluidized bed boilers,types, 224–228
Fluidized bed combustion (FBC), 215, 221, 224–230
  repowering, 228–229
FreedomCAR
  hydrogen storage system targets, 178t
  technical target performance, 176
Fresnel mirror designs, 34f
Fresnel-type concentrators, 33–34
Froth flotation. See Preconversion technology
  coal cleaning. See Column froth flotation coal cleaning
Fuel
  emissions factors/carbon intensity, 213t
  mass-transport limitation, 74–75
    degradation, 76–78
  oxidation, 192
  reactors, 254
  supply, deterioration, 228
Fuel cells
  applications, potential, 61–62
  chemical/physical processes, 192–193
  development, 65
  direct energy conversion systems, 238–239
  efficiency, 64, 68–71
  heat management, 75–76
  hydrogen utilization consideration, 191–196
  introduction, 59–64
  maximum thermodynamic efficiency, 70f
  Nernst open-circuit potential, 73
  neuron radiograph, 80f
  nomenclature, 95–97
  operating principles, 64–85
  patents, increase, 61f
  performance, 64, 68–71
  polarization curve, 72–75

illustration, 72f
reaction, 69t. *See also* Global fuel cell
reaction
references, 97–99
schematic, 65f
stacks
description, 67–68
waste heat management. *See* High-power
fuel-cell stacks
systems
advantages/limitations, 62–64
components, requirement, 67–68
technologies, summary, 194t
thermodynamic (voltaic) efficiency, 69, 71
types/descriptions/data, 66t
variant, advantages, 65, 67
varieties, 91–95
Fuel chargeable to power (FCP), 9
Fuel energy, rejection, 138–139

G

Gas boilers, 191
Gas diffusion layer (GDL), 64, 79
steam reforming, 170f
Gaseous emissions, clean-up, 243–249
Gaseous oxygen, reduction, 192
Gasification, 215, 221, 232–239. *See also*
Coal gasification
reactions, 232
reactors, types, 233
Gasification-based energy conversion options,
235f
Gasifiers, characteristics, 234t
Gas-phase reactants, activity, 68–69
Gas turbines, 136–138. *See also* Aircraft gas
turbines; Stationary gas turbines
cogeneration usage, design performance
envelope, 9f
regeneration/intercolling/reheating
equipment, 137
selection, 143–144
General Electric (GE) MS7001EA, variants,
10
Geometric concentration ratio (CR), 35
Geopressured basins, existence, 107
Geopressured fairways, 107
Geopressured resources, 106–107
Geothermal energy
applications, 108
conversion, 107–118

direct uses, 109
Geothermal fluids, reinjection, 108, 111
Geothermal heat pump (GHP), 116–118
energy-conversion process, 116
shipments, survey, 295
system. *See* Ground-loop GHP systems
Geothermal power generation (worldwide),
102t
Geothermal resources, 102–107
areas. *See* Known geothermal resource
areas
categories/types, 102–103
classification, 103t
recoverability, 103
Geothermal resources/technology
introduction, 101–102
references, 128
Geothermal resource utilization efficiency
(GRUE), 111–112
GHP. *See* Geothermal heat pump
Global fuel cell reaction, 68
Government-regulated emission standards,
213–214
Gravity separation. *See* Preconversion
technology
Gravity separators, filtered steam (expansion),
111
Greenhouse, attachment, 43–44, 45f
Gross-to-net loss, 124
Ground-loop GHP systems, 117–118
Grove, Sir William, 59
Grubb, William, 60

H

HAWT. *See* Horizontal Axis Wind Turbine
HDR. *See* Hot dry rock
Heat engine, maximum thermodynamic
efficiency, 70f
Heat generation
calculation, 76f
components, 75–76
Heat leak, 183
Heat production, rate, 31
Heat pumps. *See* Geothermal heat pump
geothermal direct use, 295t
Heat recovery steam generator (HRSG),
132–133, 140–142. *See also* Unfired
HRSG
Heat-to-power ratio, 135

Heavy hydrocarbons, partial oxidation, 168–169
Heavy metals
  clean-up, 249–250
  pollutant, 211–212
Heterotrophs, 219
High-concentration carbon dioxide steam (production), oxygen fuel combustion (usage), 252f
High Gradient Magnetic Separation (HGMS), 217–218
High heating value (HHV), 69
  comparison, 8
High-molecular-weight polymer, usage, 181
High-power density (HPD) SOFC, 88f
High-power fuel-cell stacks, waste heat management, 79
High-pressure gaseous hydrogen storage, 180–183
High-Pressure Gas Safety Institute of Japan (KHK), 183
High Pressure Regenerative Turbine Engine (HPRTE), 272–273
  combined-cycle thermodynamic processes, 272f
High surface area sorbents, usage, 190
High-temperature fuel cells, commercialization, 61
High-temperature SOFCs, power conversion efficiencies, 70
Horizontal Axis Wind Turbine (HAWT), 120
Horizontal beam, values, 27
Horizontal surface
  extraterrestrial radiation, average, 23t–24t
  mean daily solar radiation, 26f
Horizontal system, 117
Hot dry rock (HDR) geothermal resource conversion, 107f
Hot dry rock (HDR) resources, 106
Hot igneous resources, 103
Hour angle. See Earth
  definition. See Solar-hour angle
Hourly solar flux conversions, 26–27
HPD. See High-power density
HPRTE. See High Pressure Regenerative Turbine Engine
HRSG. See Heat recovery steam generator
Hybrid geothermal/fossil energy conversion, 116
Hydrogen
  atomic form, 182f

codes/standards, 199
handling process, 198–199
nature, 197–198
Hydrogen, Fuel Cells, and Infrastructure Technologies (HFCIT), 199
Hydrogen-air fuel cell, maximum thermodynamic efficiency (calculation), 70
Hydrogen burner turbines, 197
Hydrogen economy, components, 178f
Hydrogen energy
  introduction, 165–166
  references, 200–206
Hydrogen ion, physical transport, 192
Hydrogen on Demand (Millennium Cell), gravimetric capacity, 189
Hydrogen PEFC
  alternative, 78
  technical issues, 78–83
  water/heat management, 78–82
Hydrogen production, 166–176
  biological methods, usage, 176
  biomass, usage, 169–171
  partial oxidation, usage (block diagram), 168f
  steam reforming process, usage (block diagram), 167f
Hydrogen safety, 197–199
  statistics, 198t
Hydrogen storage, 176–190. See also Chemical hydrogen storage; High-pressure gaseous hydrogen storage; Liquid hydrogen storage
  alloys, consideration, 186
  carbonaceous materials, usage, 189–190
  examples, 181f
  gravimetric/volumetric storage capacities, increase (approaches), 182–183
  magnesium, capacity, 184
  material, properties, 178–179
  media, hydriding substances (theoretical capacities), 184t
  methods, types, 180f
  methods/phenomena, 177t
  options, 179–190
  theoretical volumetric hydrogen density, 185
Hydrogen utilization, 191–197
Hydrolysis reactions, 189
Hydrothermal convection systems, 102

Hydrothermal resources, 104–106
Hydrothermal resources, estimates, 110

I

IGCC. *See* Integrated gasification combined
    cycle
Illuminated $p - n$ photocell, equivalent
    circuit, 54f
Incidence angle, definition, 20f
Indirect-gain passive system, 45f
In-situ control technology, 215, 219–233
Integrated gasification combined cycle
    (IGCC), 220, 231, 233–237
    plants
        coal, usage, 235
        operating commercial scale, 237t
Intercooling, 137
Intermediate system, 145
Intermittent renewables, integration, 7–8
Internal combustion engines (ICEs), 191, 197
Internal combustion (IC) engines, 138–139
Iodine-Sulfur Cycle, 174
Ionic transport losses, reduction (SOFC
    concept), 88f

J

Joule heating, 75

K

Known geothermal resource areas (KGRAs),
    103–104, 104f
Kyoto Protocol (United Nations Framework
    Convention on Climate Change), 214,
    259

L

LAER. *See* Lowest achievable emissions rate
LaNi$_3$ metal hydride, P-C
    isotherms/Van'tHoff curve, 186f
Legislated efficiency, 148
Levelized cost of electricity (levelized COE),
    1
    calculation, 4–5
        direct-unit basis, 5
    hours, contrast, 6
Lift principle, usage, 120
Lignite, characteristics, 209
Liquidated air-based flat-plat collectors,
    cross-section, 30f

Liquid-based space-heating system, schematic
    diagram, 42f
Liquid-cooled flat-plate collectors, test loop
    (usage), 38–39
Liquid-dominated resources, 104, 105–106
    system selection considerations, 115–116
Liquid hydrogen (LH$_2$) storage, 183
Load duration curve, 6f
Load following, 145
Local standard time (LST), relationship, 16
Los Alamos National Laboratory, PEFC
    research, 60–61
Lowest achievable emissions rate (LAER),
    151–152
Low-grade coals IGCC, heat rate level, 236
Low heating value (LHV), 69
    comparison, 8
Low NO$_x$ burners (LNBs), 215
    illustration, 223f
    usage, 223

M

Magma resources, 106
Magnetohydrodynamics (MHD), 221, 232
Marrakesh Accords, 259
MCFC. *See* Molten carbonate fuel cell
Mechanical solar space heating system, 42–43
Membrane electrode assembly (MEA), 84
Membrane scrubbing. *See* Carbon dioxide
Mercury
    presence, 249
    removal. *See* Elemental mercury
Metal/complex hydrides, usage, 183–188
MHD. *See* Magnetohydrodynamics
Microbial fuel cell (MFC), 95
Molten carbonate fuel cell (MCFC), 59,
    91–95
    advantages/disadvantages, 93–94
    relative performance, 93f
    temperature operation, 92–93
Monolithic SOFC design, schematic, 89f
Monthly averaged solar flux conversions,
    27–28
Multipass system, usage, 231

N

Nafion, 172
    conductivity, plot, 79–80
    electrolyte conductivity, 80f

National Ambient Air Quality Standards
(NAAQS), 150, 152
National Climatic Center (NCC), 25
National Energy Act, 147
National Fire Protection Association (NFPA),
199
National Hydrogen Association (NHA),
Codes and Standards Working Group
(creation), 199
National Renewable Energy Laboratory
(NREL), 199
Natural gas, steam reforming, 166–168
Nernst equation, usage, 68
Net geothermal brine effectiveness, 115f
Net heat to process (NHP), calculation, 9
NETL. *See* U.S. National Energy Technology
Laboratory
New source performance standards (NSPS),
150
New source review (NSR), 151–152
NFPA. *See* National Fire Protection
Association
NHA. *See* National Hydrogen Association
Nitrogen
elimination, 253
oxidation, 247
Nitrogen oxides (NO_x)
clean-up, 246–248
manmade sources, 211f
pollutant, 210
problems, 247
reburning, 224
Nitrous oxide, clean-up, 248–249
Noble-metal catalyst loading, 61
Nonattainment (NA) program, 151
Nonattainment (NA) rate, 154
Noncondensing steam turbines, 136
Nonconvecting solar pond, schematic
diagram, 46f
Nondispatchable fuel resources, 8
NO_x Budget Trading Program (NBP), 3
NREL. *See* National Renewable Energy
Laboratory
NSPS. *See* New source performance standards
NSR. *See* New source review

O

Obligated autotrophs, 219
Office of Science and Engineering Research
(OSER), 196
Off-normal solar radiation, intensity, 19–20

Ohmic loss region, absence, 73
Open-circuit voltage (OCV), achievement, 72
Open-loop system, 117–118
Operations and management (O&M) costs, 6
ORR kinetics, efficiency, 91
Ortho-para conversion, 183
OSER. *See* Office of Science and Engineering
Research
Output-based format, 152
Output/insolation input ratio, 54, 56
Oxidizer, mass-transport limitation, 74–75
Oxy-fuel combustion, usage, 253

P

PAFC. *See* Phosphoric acid fuel cell
Parabolic trough, 34f
Paraboloidal dishes
construction, 36
segmented mirror approximation, 36f
Partial gasification combined cycle (PGCC),
238
plants, 220
Partial oxidation (POX), 168. *See also* Heavy
hydrocarbons
Particulates
clean-up, 239–243
pollutant, 211
removal, 236
Passive solar space heating system, 43–44
types, 43–44
Peaking system, 145
PEC. *See* Photoelectrochemical
PEFC. *See* Polymer electrolyte fuel cell
PEM. *See* Proton exchange membrane
Perfluorocarbon ion exchange membranes,
172
Perfluorosulfonic acid-polytetrafluoroethylene
(PTFE), 79
PFBC. *See* Pressurized fluidized bed
combustor
PGCC. *See* Partial gasification combined
cycle
Phosphoric acid fuel cell (PAFC), 60, 94–95,
195
Photoelectrochemical (PEC) cell, schematic,
175f
Photoelectrochemical (PEC) hydrogen
production, 174–176
Photovoltaic (PV) cells/modules, 277–294

annual shipments, 287t, 288t
distribution, 288t
export shipments, 291t
destination, 291t–293t
shipments, 279t–281t, 289t, 290t
values, 289t
Photovoltaic (PV) collectors, 275
Photovoltaic (PV) converters, maximum
theoretical efficiency, 53f
Photovoltaic (PV) domestic shipments, 287t
Photovoltaic (PV) exports, 283
Photovoltaic (PV) manufacturing industry,
employment, 294t
Photovoltaic (PV) solar energy applications,
49–56
Photovoltaic (PV) systems
electrical storage/control system,
requirement, 54
shipments, 294t
Photovoltaic-related (PV-related) activities,
company involvement, 294t
Physiochemical degradation, modes, 76–78
Plant emissions, reduction, 228
Polarization. *See* Activation polarization
curve. *See* Fuel cells
linear region, evidence, 74
Polybenzimidazole (PBI), 172
Polyetheretherketone (PEEK), 172
Polyethersulfone (PES), 172
Polymer electrolyte fuel cell (PEFC), 60. *See
also* Hydrogen PEFC
durability, 82–83
freeze-thaw cycling, 82–83
operating efficiency, 75
water transport/generation, schematic, 81f
Polyphenylquinoxaline (PPQ), 172
Pond/lack system, 117
Postcombustion CO_2 extraction, 255–256
Postconversion technologies, 215
clean-up, 239–251
Power
environmentally friendly sources, factors,
3–4
production, CHP systems (comparison),
269f
Power generation technologies
economic comparisons, 1
economic evaluation, 4–7
introduction, 1
market growth/emissions, 2–4
references, 11

Power plant waste heat, application, 270–271
Preconversion technology, 215–219
agglomeration, 217
biological cleaning, 219
chemical cleaning, 218
froth flotation, 217
gravity separation, 216–217
physical cleaning, 216–218
Pressure-composition isotherms, usage, 185
Pressure swing absorption (PSA), 255
Pressurized fluidized bed combustion plants,
220
Pressurized fluidized bed combustor (PFBC),
224, 229–230
Prevention of significant deterioration (PSD),
151–152
Price duration curve, 7f
Prime movers
descriptions, 135–139
inclusion, 135
selection/sizing, 143–144
Proton (polymer) exchange membrane (PEM),
172
fuel cells, 195
PSA. *See* Pressure swing absorption
PSD. *See* Prevention of significant
deterioration
PTFE. *See* Perfluorosulfonic
acid-polytetrafluoroethylene
Public Utility Regulatory Policies Act
(PURPA), 147–149
Pulverized coal-fired furnaces, flame
temperatures, 222–223
Pulverized coal (PC) boiler
LNB usage, 221
tools, 231
Pulverized coal (PC) combustion, 219–220
process, 221–224
Pyranometer, usage, 25f
Pyrolysis, 170–171, 232

Q

Qualified facility (QF), 148
efficiency standards, requirement, 148t
Quantitative solar flux availability, 22–28

R

Radiation, ambient temperature (usage), 36
Radiative heat loss, reduction, 32

Rankine cycle, 134
Brayton cycle, combination, 229
plants, supercritical steam cycle, 220
steam plants, temperatures, 222
thermodynamic representation, 229f
Rankine steam power plants, fossil fuel
burning, 267–268
Reasonable available control technology
(RACT), 151
Receiver, infrared emittance, 36
Reciprocating engines, 138–139
power ratings, 138
Regeneration (reoperation) process, 137
Reheating, 137
Reversible decomposition potential, 171
Riffled tubes, usage, 231
Rotating drum coal washing, 216f
Rotor RPM, conversion, 125

S

SCR. *See* Selective catalytic reduction
SECA. *See* Department of Energy Solid State
Energy Conversion Alliance
Segmented cell-in-series design, schematic,
90f
Selective catalytic reduction (SCR), 215,
247–248, 248f
Selective noncatalytic reduction (SNCR), 215,
248, 248f
Selective surfaces, 32–33
properties, 33t
Sensitivity analyses, 158
Sequestration. *See* Carbon dioxide
Silicon cell, current-voltage (IV)
characteristics, 55f
Single-curvature solar concentrators, 34f
Single-tank indirect solar water-heating
system, 40f
Sloshing, 183
SMR. *See* Steam methane reformation
SNCR. *See* Selective noncatalytic reduction
SOFC. *See* Solid oxide fuel cell
Solar-altitude angle
diagram, 17f
relationship, 16
Solar and Energy Conversion Laboratory
(University of Florida), 273
Solar-azimuth angle
calculation, 16
diagram, 17f

Solar cells, proportional production, 53–54
Solar collectors
closed-loop testing configuration, 38f
movement, 20–22
schematic diagram, 30f
Solar declination, 15f
Solar energy
applications, 13
availability, 13–28
references, 56–57
utilization, 175
Solar flux. *See* Extraterrestrial solar flux;
Terrestrial solar flux
availability. *See* Quantitative solar flux
availability
conversions. *See* Daily solar flux
conversions; Hourly solar flux
conversions; Monthly averaged solar
flux conversions
daily absorption, 50f
Solar fraction, empirical equations, 50t
development, 49
symbols, definitions, 51t–52t
Solar geometry, 13–17
Solar-hour angle, definition, 15f
Solar One, 47
Solar ponds, 44–47
installation, requirements, 46–47
Solar position, 16–17
example, 18t
Solar-produced power, cost effectiveness,
47–48
Solar space heating systems. *See* Mechanical
solar space heating systems; Passive
solar space heating system
Solar thermal applications, 39–49
Solar thermal collectors, 275–277
activities, company involvement, 286t
annual shipments, 277t, 278t
exports, distribution, 282t
industry, employment, 286t
shipments, 279t–281t, 284t
distribution, 283t
percentage, 285t
systems, shipments, 285t
usage, 28–39
Solar thermal power production, 47–48
Solar thermal processes, performance
prediction, 48–49

Solar utilizability, 49

Solar water heating, 39–41

    passive thermosiphon single-tank direct system, 41f

Solid electrolyte fuel cells, 195–196

Solid oxide fuel cell (SOFC), 59, 85–91, 195–196

    advantages, 85

    design, schematic. *See* Monolithic SOFC design

    durability, 86

    flow channel material structure, 87

    fuel cell operation, 238f

    lower temperature operation, desire, 86

    monolithic/segmented cell-in-series designs, 89

    performance/materials, 86–91

    sealless tubular concept, drawback, 87, 89

    sealless tubular design, 88f

    technical issues, 86

Solid polymer electrolyte (SPE) electrolyzer, 172

    schematic, 172f

Solid waste, clean-up, 250–251

South Coast Air Quality Management District (SCAQMD), 151

Spiral tube arrangement, usage, 231

Spray-dry scrubbing

    process, 246f

    usage, 245

State Implementation Plans (SIPs), 150

Stationary (industrial) gas turbines, 136–137

Stationary power applications, 62

Stationary power sources, installation projects (estimation), 63f

Steam-based Rankine-cycle electric-power-generation facilities, 152

Steam conversion. *See* Direct steam conversion; Flashed steam conversion

Steam methane reformation (SMR), 166

    process, expense (comparison), 168

    usage, 167

Steam turbines, 136. *See also* Noncondensing steam turbines

    selection, 143

Stefan-Boltzmann constant, usage, 36

Sulphur capture, 225

    process, 226f

Sulphur dioxide (SO_2)

    clean-up, 244–246

    manmade sources, 211f

    pollutant, 210–211

Sunlight, photovoltaic conversion, 50, 53

Sun path

    solstice/equinox, 17

    summer solstice/equinox, 19f

    winter solstice, 19f

Sunrise

    hour angle, 28

    occurrence, 17–19

Sunset, occurrence, 17–19

Sunspace

    attachment, 43–44

    passive heating system, 45f

Supercritical boilers, 215, 230–231

    technology, 221

    water/steam flow, once-through type (usage), 230–231

Supercritical power plant, steam condition/efficiency, 230f

Support fuel, cost reduction, 228

Surface filtration, 239–241

Syngas, 233

T

Terrestrial solar flux, 25

Thermal applications, 48

Thermal energy, storage, 48

Thermal storage wall (TSW)

    system, 45f

    usage, 43

Thermochemical cycle, schematic diagram, 173f

Thermochemical hydrogen production, 173–174

Thermoelectric power plant, example, 271–272

Thermosiphon approach, 41

Topping cycles, 132–133

Total COE, production/effectiveness, 6

Total radiation, beam component (locating), 26–27

Tracking collectors, solar incidence angle equation, 21t

Tropics, location, 14f

Trough collectors, 33–35

Turbines. *See* Gas turbines; Hydrogen burner turbines; Steam turbines; Wind turbines

Turn-key system, purchase, 146–147

Two-stage flash conversion, 113f

U

Uncontrolled NO_x emissions, 247
Underground/undersea sequestration. *See* Carbon dioxide
Unfired HRSG
  heat transfer, temperature function, 141f
  schematic, 141f
U.S. Acid Rain NO_x Reduction Program, 2
U.S. Geological Survey (USGS), geothermal resources assessment, 103
U.S. Geothermal Resource Base, 103–104
U.S. National Energy Technology Laboratory (NETL), 196, 259
U.S. National Weather Service (NWS), solar flux data, 25
U.S. wind turbine installations, 120f
US-DOE hydrogen storage milestones, 177t
UT-3 Cycle, 174

V

Vapor-dominated resources, 104, 105
Vehicular hydrogen storage, options, 179f
Vertical Axis Wind Turbines (VAWT), 120
Vertical system, 117
Voltage/temperature, change, 56f
Volumetric/gravimetric hydrogen density, 185f

W

Waste heat
  application. *See* Power plant waste heat
  usage, 267
    references, 274
Water balance, achievement, 82

Water heaters, circulating pumps (exclusion), 41
Water-heating systems, usage, 41
Water Quality Act of 1965, 153
Wet FGD process, 245f
  usage, 244–245
Wet scrubbers
  types, 243f
  usage, 243
Wind power generation, 119
  configurations, 120–121
  energy yield, 121–124
  market/economics, 119–120
  peak efficiency, 122
  power equation, 121–123
  power production, 121–124
  references, 128
  rotor/drive train design, 124–126
  site selection, 126–128
Wind speed probability distribution, notional power curves, 123f
Wind turbines
  boundary layer impact, power-law equation, 127
  performance, 123
  placement, micrositing (usage), 126–127
  rotor/drive train design, 124–126
World electricity share projections, fuel source basis, 2f

Z

ZnO/Zn Cycle, 173–174